CAMBRIDGE COUNTY GEOGRAPHIES

General Editor: F. H. H. GUILLEMARD, M.A., M.D.

T0352221

LINCOLNSHIRE

CAMBRIDGE UNIVERSITY PRESS
Cambridge, New York, Melbourne, Madrid, Cape Town,
Singapore, São Paulo, Delhi, Mexico City

Cambridge University Press
The Edinburgh Building, Cambridge CB2 8RU, UK

Published in the United States of America by Cambridge University Press, New York

www.cambridge.org
Information on this title: www.cambridge.org/9781107612648

First published 1913
First paperback edition 2013

A catalogue record for this publication is available from the British Library

ISBN 978-1-107-61264-8 Paperback

Cambridge County Geographies

LINCOLNSHIRE

by

E. MANSEL SYMPSON,
M.A., M.D., F.S.A.

With Maps, Diagrams and Illustrations

Cambridge :

at the University Press

1913

TO MY WIFE

CONTENTS

ILLUSTRATIONS

MAPS

Thanks are due to Messrs Clayton and Shuttleworth for the illustration on p. 7; to the late Mr F. M. Burton for those on pp. 25 and 72; to the Lincoln Museum for those on pp. 27, 59 and 113; to Mr N. Sutton-Nelthorpe for those on pp. 39 and 68; to Messrs Ruston Proctor and Co. and Messrs Marshall, Sons and Co. for those on pp. 80 and 81 respectively; to the Great Central Railway Co. for that on p. 92; to the Rev. A. Hunt for that on p. 101; to Mr Aymer Vallance for that on p. 129; to the Rev. Canon Cole for that on p. 146; to Mr A. H. Tod for that on p. 167; to the Rev. G. E. Mahon for that on p. 174 and to Dr J. S. Chater for that on p. 180.

1. County and Shire. The Word *Lincolnshire.*

The word county is derived from an old French word meaning a province governed by a count (French), or an earl (Saxon) as he was afterwards termed, and it is applied generally to all the provinces in England, particularly to those which had been separate kingdoms in Anglo-Saxon times, like Kent.

The last syllable of the word Lincolnshire is of Anglo-Saxon origin, and means a portion of a state or kingdom—a part *shorn* or cut off—by this reminding us that Lincolnshire was once one of the provinces of the great mid-England kingdom of Mercia. It is interesting to note that nearly all the shires into which this kingdom was divided have taken their names from their capital towns or cities, i.e. *Leicester*shire. Out of 17 shires thus formed the only real exception is Rutland, for Cheshire is probably corrupted from *Chester*shire and Shropshire from Scrobbesbury or Shrewsbury.

The sheriff—the shire-reeve or steward—is the king's officer and his representative in the shire. Besides a sheriff for the county, Lincoln City has one for itself, having been given the dignity of a county in itself by King

Henry IV. The first part of the name of the county is of course taken from the capital city of Lincolnshire

Lincoln from the South-east

—Lincoln. This name, like the city to which it belongs, has a very ancient origin. By the British the town was called Lindon (*Llyn* being still the Welsh for a lake, and *dun* or *don* for a hill-fort), meaning "the hill-fort by the water." Ptolemy, writing about the year A.D. 120, says that the two chief towns of the Coritani—a British tribe which inhabited the present counties of Lincolnshire, Rutland, Leicestershire, and part of Nottinghamshire, Warwickshire, and Derbyshire—were Lindum (a Latinized form of the name) and Ragae (Leicester). After the Roman conquest of this part of Britain Lindum became a Roman fortress and later a colony, with the title Lindum, or *Lindi colonia*. By the Saxons it was called Lindecollinam, or Lindocyllanceaster (*ceaster* or *chester* generally indicating Roman stone fortifications). By the Normans writing in Latin it was named Lincolnia, but in Norman-French Nicole (a curious instance of the transposition of letters) as late as the time of King Edward IV, and the same applied to the name of the county, called by the Normans Nicoleshire, though Domesday Book calls it Lincolescire. The division of this country into counties or shires probably began before the time of King Alfred, and was completed when the kingdom of King Edgar brought all divisions of the people under one rule. Lincolnshire would thus be separated off from Mercia, which got its name from bordering on the Marches or frontiers of Wales.

Lincolnshire was, possibly from about the same time, subdivided into three, called The Parts of Lindsey, The Parts of Kesteven, and The Parts of Holland. Of these the first, Lindsey (in Domesday Book Lindesie), gets its name either from Lincoln, or as being the *eye* or

island of the Lindissi who inhabited it. It occupies about
the northern half of the county, and according to Domesday Book, it again was subdivided into three Redings or
Ridings (i.e. thirdings)—North, South, and West. The
second division, Kesteven (in Domesday Chetsteven), takes
the south-western quarter, and its name is believed to be
derived from Coedstefne, "the wood jutting out into the
fen." The third, Holland (in Domesday Hoiland), takes
the remaining south-eastern quarter, and signifies, as does
its namesake across the North Sea, "the hollow land," as
being below the level of the sea in many parts.

2. General Characteristics.

Lincolnshire—situated on the east coast of England
a little north of its middle point—is a maritime county
with a most important estuary on the north, the Humber,
wherein is her greatest port, Great Grimsby, possibly the
largest fishing port in the world, and a lesser and much
shallower one on the south, the Wash, where is her lesser
port, Boston. These two ports furnish practically all the
harbours she has along her 85 miles of coast, or 117 if
we measure from the junction of the Trent, Ouse, and
Humber to the beginning of the Norfolk coast. Flat
and low though that coast may be, guarded by "a sand-
built ridge of heaped hills that mind the sea," still there
is a fascination in the great extent of yellow sands exposed
by her shallow seas, just as there is a recompense for the
level plain of marsh or fen in the vast expanse of sky

where dawn or sunset are seen at their finest. As Charles Kingsley says, speaking of these very fens, in *Hereward the Wake*, "They have a beauty of their own, these great fens, even now when they are dyked and drained, tilled and fenced—a beauty as of the sea, of boundless expanse and freedom." And after an exquisite appreciation of

The Fens in Winter: Cowbit Wash, near Spalding

the fen in Norman times he adds, "Overhead the arch of heaven spreads more ample than elsewhere, as over the open sea ; and that vastness gave, and still gives, such cloudlands, such sunrises, such sunsets, as can be seen nowhere else within these isles." And the views are indeed marvellous ; from Alkborough over the junction of the Trent and Ouse to the Humber, from the Wolds above Caistor westwards towards Lincoln, from Lincoln eastwards to those same Wolds, or westwards to the hills of Nottinghamshire and the mountains of Derbyshire, or looking eastwards from the edge of the " high Wold " over the great plain of marsh,

> " That sweeps with all its autumn bowers,
> And crowded farms, and lessening towers
> To mingle with the bounding main."

Lincolnshire is pre-eminently an agricultural county. Camden, writing in 1590, describes the county as of " a most mild climate fit for corn and cattle, adorned with numerous towns, and watered by several rivers." The farms on the Heath, the Cliff, the Wolds, and the Fens are examples of first-rate cultivation of the soil in corn and roots, while the neighbourhood of Boston is well known for its potatoes and growth of mustard for seed, and the Isle of Axholme for vegetables. The Marsh furnishes one of the very best grazing lands in the country. The Lincolnshire red shorthorn forms a distinct and valuable kind of cattle, the Lincolnshire sheep is of large size with a heavy fleece, while the reputation of its horses at Horncastle and Lincoln fairs has been excellent for many years.

Lincolnshire is also a great industrial county. The engineering works at Lincoln, Gainsborough, and Grantham, giving employment to many thousands of workmen, are famed for all kinds of agricultural machinery, as well as for that used in milling and mining. At Scunthorpe are large and important iron mines and smelting works. Steam mills for corn surround Brayford Pool in Lincoln, where, as at Gainsborough, are large cake mills and wood-works, while Sleaford has the largest malt-houses in the country.

Iron Works, Lincoln

3. Size. Shape. Boundaries.

Lincolnshire is the second largest county in England, its area comprising 1,705,293 acres. It takes position between Yorkshire (with the enormous acreage of 3,889,758 acres) and Devonshire (with an acreage of 1,671,364), and it has a population of 563,960.

John Speed, writing in 1627, quaintly remarks: "The forme of this County doth somewhat resemble the body of a Lute, whose East Coasts lye bowe-like into the German Ocean." Fuller, in 1662, compares it to "a bended bow, the sea making the back, the rivers Welland and Humber the two horns thereof, whilst Trent hangeth down like a broken string, as being somewhat of the shortest." The shape of Lincolnshire might also be likened to a pear, with one-third of its western side sliced off, and with the tail turned up to join Norfolk and form part of the southern boundary of the Wash.

Measuring from the northernmost projection into the Humber to the southernmost part of the boundary at Stamford its length is about 75 miles, and its breadth from the easternmost point at Ingoldmells on the east coast to the Nottinghamshire county border at North Scarle on the west is about 47 miles.

Its boundaries are natural ones for the greater part. It is bounded on the north by the river Humber; on the east by the North Sea; on the south by various drains and the river Welland dividing it from Norfolk, Cambridgeshire, and Northamptonshire; and on the west by an artificial border-line between it and Rutland, Leicestershire, and Nottinghamshire. Higher up, the river Trent bounds West Lindsey, while the Isle of Axholme is separated from Yorkshire chiefly by an artificial and exceedingly irregular boundary line. This part of the county appears, on the map, as if it must have been captured for Lincolnshire from the old royal

Junction of the Trent, Ouse, and Humber

(*View looking Westward from near Alkborough*)

forest of Hatfield Chase in Yorkshire, as it is on the
west side of what, obviously, is *the* natural boundary.
On the other hand Nottinghamshire appears to "cut a
monstrous cantle out" of Lincolnshire on the east side
of the Trent between Newark and Newton-on-Trent,
of which the north-easternmost point actually comes
within five miles of Lincoln. Some authorities have
believed that Newark and its wapentake were wrenched
away from Lincolnshire about the time of King Alfred,
and, in this connection, it is interesting to notice the
fact that, in the later years of Remigius, first Norman
Bishop of Lincoln, the Archbishop of York claimed a
large part of the provinces of Lindsey, with Lincoln,
Louth, Stow, and Newark, as properly subject to his
sway.

4. Surface and General Features.

The surface of our county is marked by two long
lines of hills, the Cliff (or Heath) and the Wolds. There
is also a vast tract of Fen, a word now used to denote
land which is or has been overflowed by fresh water, and
a wide belt of Marsh, a term in Lincolnshire applied to
land which has been or is liable to be overflowed by
sea-water.

The Cliff, which is of oolite limestone, reaches from
Winteringham on the Humber, in a course almost due
south, to Grantham and the southern limits of Lincolnshire. It has a general elevation of some 200 feet, and

has a fairly sharp escarpment rising 100 feet or more above the western plain. For most of the distance this western plain is the valley of the Trent, which has but low banks, save those of New Red Sandstone at Gainsborough and Newton, on the eastern or right bank.

The Cliff is interrupted by the great gap at Lincoln, and again by the lesser one at Ancaster, where its height has risen to nearly 400 feet. It is here called "the Heath," and its western borders are known as the Vale of Belvoir. Broadening out before reaching Grantham it attains, near Wyville and Buckminster, its greatest elevation of 500 feet. The little Witham valley is picturesque and well wooded by Easton and North Stoke Rochford Parks. This range of the Cliff is really the edge of an inclined tableland which slopes gently eastward, in Lindsey to meet the rising Wolds, and in Kesteven to the Fens, the parts between Grantham, Sleaford, Bourne and Stamford being particularly well wooded and picturesque.

Another range of Cliff is found on the right bank of the Trent at Alkborough; here it is about 200 feet high, with woods sloping steeply down to the river. The formation is of the New Red Sandstone. Much of this part of the county was a sandy waste, given up to rabbit warrens. At Scawby gull-ponds the scenery might pass for Scottish, from the sandy undulations round the lake and the abundance of pine trees.

The chalk Wolds stretch from Barton-on-Humber and South Ferriby in a south-easterly direction for nearly 50 miles, with an average width of about eight miles, widening as they proceed, to Burgh and Spilsby. The highest point is probably just above Normanby, 548 feet, though a fair acreage is above the 400 feet level. They consist of great rolling downs, intersected with deep valleys. Near Brocklesby they are well wooded, as Pelham Pillar was erected to record the planting of 12,552,700 trees between the years 1787 and 1823 by the Lord Yarborough of that date. In the neighbourhood of Somersby, Tennyson's birth-place, there is a fair amount of wooded country, and one of the most charming valleys, Well Vale, is close to Alford. The farms on the Wolds are of great size; possibly the largest is at Withcall, which runs to 2750 acres.

Three miles south-east of Lincoln, at the village of Washingborough, begin the Fens, marked by the black peaty soil[1]. Hence they stretch southwards and south-eastwards to Boston, Spalding, and Cambridge. A little piece of true unreclaimed fen remains near Dogdyke, but all the rest are drained, almost too well, as they are sometimes short of water in the summer. The numerous dykes take the place of hedges, which are comparatively rare. Wheat, beans, and roots all grow finely on this soil, and near Boston potatoes and mustard grown for seed. The names along the Witham ending in *ey*, such as Bardn*ey* and Southr*ey*, show the islands which alone could have been inhabited amidst the surrounding swamp in the pre-drainage days.

The Marsh is a strip of land between the Wolds and

---

[1] Occasionally in dry seasons this underlying peat takes fire and is difficult to put out, as happened on the edge of a dyke in Branston Fen in August, 1911.

A Wold Farm

the North Sea. It is alluvial in origin, its width varies
from six to 10 miles, and it reaches from below Grimsby
to Skegness. It is protected from the sea only by sand-
dunes, covered with swordgrass and the spiky sallow-thorn,
with its grey green leaves, and berries of a warm orange
colour in autumn. At Marsh Chapel and North Somercotes, it has been enlarged by the draining and capture
from the sea of the salt fitties (this being an old Norse
word for the out-marshes between the sea-bank and the
sea). This is Tennyson's

> "Waste enormous marsh
> Where from the frequent bridge
> The trenched waters run from sky to sky."

Much of it is first-rate grazing ground for sheep and
cattle, and rich crops of wheat, large eared and long
stemmed, are raised.

The Isle of Axholme, almost a dead level, has been
carefully drained, and is enriched by warping (which will
be further alluded to in a later portion of this book). The
hills round Epworth contain gypsum and belong to the
New Red Sandstone formation.

5.   Rivers and Watersheds.

The rivers of this county, with one exception—the
Trent—are small and slow. Macaulay poured righteous
scorn on the line "Streams meander level with their
fount" as an impossible feat, but the Witham comes a
little near the poet's description between Lincoln and

Boston, as the level of Brayford is only 16 feet above the sea. In their earlier miles some of the rivers are lively streams, and afford good trout fishing; to their more sluggish later portions come thousands weekly, for coarse fishing, from the large industrial cities of Yorkshire. The celebrity of the Witham for pike has been hymned by Spenser and Drayton, and there was an old and popular saying,

> Wytham eel and Ancum [Ancholme] pike
> In all the world there is none syke.

We will begin with those rivers which arise within the county. There is a watershed on the Cliff, about three miles south of the cross-road between Gainsborough and Market Rasen, and about a mile eastwards of the Ermine Street, the great North Road from Lincoln to the Humber. The river Ancholme from the village of Spridlington is joined at Bishop's Bridge by the little river Rase, which rises in the heart of the Wolds at Bully Hill, and runs west past Tealby and Bayons Manor, the handsome modern castle of the Tennyson D'Eyncourts, through the three Rasens—Market, Middle, and West— to which it gives its name. From Bishop's Bridge the Ancholme proceeds almost due north, through low-lying fields called carrs (car = fen), to fall into the Humber at Ferriby Sluice, Brigg (properly Glanford Brigg) being the only place of any importance on its way. It thus serves the useful purpose of draining the valley between the Cliff on the west and the Wolds on the east. For its last 16 miles a new straight course has been cut, called

the New Ancholme river, which for five miles north of
Brigg is accompanied by another dyke, the Weir Dyke.
Continual complaints of the damage done to the drainage
of the Ancholme valley seem to have been made from
the time of King Edward I onwards, and in the 21st year
of Charles I's reign Sir John Monson undertook to
drain this area. This apparently was well done, but
spoilt during the great civil war—according to the account
given by Dugdale—by the works being neglected, the
drains filled up, and the sluices damaged. Improvements
were made in the reign of King George III, and others
about the year 1826 and afterwards.

Rising from one source in Hackthorn within a mile or
two of that of the Ancholme, and from another in Busling-
thorpe, the Langworth river, about nine miles long, runs
south past Snarford, through Langworth (having got
accessory streams from Welton and Dunholme, and from
Riseholme and Sudbrook lakes, and others from the
western slopes of the Wolds), past the site of the once
powerful and important Abbey of Barlings, to join the
Witham, about seven miles east of Lincoln, close to a
pumping station called Short Ferry. In the village of
Ludford, six miles east of Market Rasen, the river
Bain has its source, and for some 13 or 14 miles runs
nearly south to Horncastle, part of whose Roman name,
*Bano*vallum, came from the stream. It passes *en route*
the picturesque villages of Burgh and Donington, and the
parks of Girsby and Biscathorpe. From Horncastle, for
the 10 miles to the Witham, an attempt was made at the
beginning of the last century to render it navigable. Its

course is south with a slight curve to the west, passing
near to Scrivelsby Court (the home of the Champions of
England, the Dymokes), Kirkby, the woods of Tumby,
the parish and village of Coningsby, with its fine church
tower, and Tattershall church and castle.   The Bain falls
into the Witham at Dogdyke (originally *Dock* Dyke).
The Horncastle Canal, which leaves it at Tattershall,
joins the Witham about two miles farther east.   The

Tetney Lock

other main watersheds of the county are formed by the
line of the chalk Wolds extending from the Humber at
Ferriby south-eastwards for 50 miles as far as Spilsby,
and by the line of oolite limestone Cliff running almost
due north and south from the Humber at Winteringham
to below Grantham.

The streams to the north and east of the Wolds are
not of much importance.   The river Lud (which gives

its name to Louth) rises in Withcall, runs through Louth, and by a devious course past the site of Louth Abbey, Alvingham, Conisholme, and North Somercotes, finds its way into the sea a little north of Donna Hook coastguard station, at Grainthorpe Haven. About three miles north of this opening is Tetney Haven, where the Louth Navigation canal—made in 1763 from Louth—joins the sea, having left the river Lud at Alvingham. Through the marsh lying between the Wolds and the sea flow various other streams which generally have been dealt with as land drains, carefully embanked and their course straightened, as they approach the coast. The river Steeping, at its source called Lynn, runs south-eastwards between two spurs of the Wolds, one ending at Spilsby, and the other at Welton-le-Marsh, and reaches the sea at Gibraltar Point.

The same line of east to west watershed holds good, as was noticed when speaking of the Ancholme, with regard to the streams on the western side of the Cliff from the Humber to Lincoln. North of a line from Market Rasen to Gainsborough, these, like the river Eau and Bottesford Beck, run westwards into the Trent, or, like Winterton Beck, northwards into the Humber. Below this line, at Corringham, is the origin of the little river Till, which proceeding southwards and eastwards joins the Fossdyke (the Roman canal between Lincoln and the Trent) at a point about four miles west of Lincoln, where is a farm with the Scandinavian-sounding name of Odde or Hadde.

The Witham is the largest and most important of the

purely Lincolnshire rivers, being about 70 miles in length, and draining on its way to Lincoln the flat country between the edge of the Cliff and the Trent, and on its way to Boston the great expanse of fen, altogether about 1070 square miles, of which 414,988 acres are highlands, and 265,404 fen land. Its course may be compared to the shape of a horseshoe, one end of which is at South Witham, 10 miles south of Grantham, whence it starts, the middle of the rounded top at Lincoln, and the other end at Boston, where it joins the sea. After leaving South Witham, it passes between Woolsthorpe (in whose manor house Sir Isaac Newton was born) and Colsterworth, between the two fine Elizabethan houses, Easton and Stoke Rochford, and past Great Ponton, whose striking west tower, built in 1519, must be well known to the passengers by the Great Northern line. After leaving Grantham it proceeds northwards, skirting the pleasant woods of Belton and Syston, till it is confronted with the hill on which Hough stands, and whence the little river Brant starts. Here it turns westwards to Westborough, near which it is joined by the Foston Beck, coming from Denton reservoir, and at Long Bennington bends northwards again to Claypole, where a fourteenth century bridge was destroyed by a Rural District Council in 1905. Just by Hough may be noticed a sudden drop in the Cliff of some 200 feet, leaving a gap leading through to Ancaster and Sleaford. It is quite likely that the original course of the Witham here turned eastwards through this gap and so to the sea. Passing from Claypole by Barnby, Norton Disney (the home of the

St Vincents) and Aubourn (not "the loveliest village of the plain," but still quite sufficiently attractive), it is joined by the Brant and goes northwards to Lincoln, where it runs into Brayford, the pool below the hill (representing a huge lake or morass of earlier times), which is joined to the Trent by the Fossdyke, the canal made by the Romans and restored by Henry I. From Brayford it turns sharply to the right, flows through the city of Lincoln, passing beneath the medieval High Bridge, with its picturesque half-timbered houses on the western side, and, practically canalised, runs for about 30 miles through the Fen to its outlet in Boston Deeps, passing the sites of the once important monasteries of Bardney and Kirkstead, the castle of Tattershall, and Boston with its splendid church and noble tower. It is joined on its way by many streams, of which one, the Sleaford river, runs—as Kyme Eau—to Chapel Hill. Possibly another route for the Witham was across from Dogdyke to Wainfleet; but at all events by 1240 it must have run to Boston, as there are extant notices as to its banks. The tide flowed as far as Lincoln, raising the water at Swanpool two feet, till in the year 1500 Mayhowe Hake, an engineer of Gravelines, was ordered to make a sluice at Boston to stop it. Another sluice was made in 1543 at Langrick, and there are now others at Bardney and Lincoln.

The Glen is about 31 miles long and rises at Somerby, near Grantham. Passing southwards it skirts Bourne and passes through Deeping Fen, through Pinchbeck, north of Spalding and Surfleet, to join the Welland at the Reservoir.

High Bridge, Lincoln

The Welland enters Lincolnshire at Stamford, runs
through Market Deeping between Peakirk (where St
Guthlac's sister had a cell) and Deeping St James, and
thence past Crowland Abbey and Spalding, to fall into
the Wash at Fossdyke.

The river Nene only belongs to this county from

The Welland at Spalding

Tydd St Mary's to its entry into the Wash, having passed
under Sutton bridge. The scene of King John's disaster
with his baggage is between the bridge and the Wash.

The Trent is, of course, far the most important river
which is connected with the county and does not rise in
it. Its source is on the northern border of Staffordshire,
whence it flows by Derbyshire, Leicestershire, and

Nottinghamshire to our western border. Its course is almost directly northwards from Newark, and it begins to form the western boundary of Lincolnshire just south of Dunham Bridge and Newton. The cliffs here, of New Red Sandstone, are very picturesque, and much picnicking goes on during the summer. Some four or

Torksey Bridge

five miles further down the river is Torksey, the ancient Tiovulfingcester, where the Fossdyke joins the Trent. Further still on the right is Marton, with the Roman road, Tillbridge Lane, and Littleborough, the Roman Segelocum, on the left. Some picturesque windings lead past Knaith, the birthplace of Thomas Sutton, founder of the Charterhouse, and Lea, the home of our distinguished

antiquary, the late Sir Charles Anderson, to Gainsborough
(the St Ogg's of *The Mill on the Floss*). Here the right
bank is high and wooded, and has a Danish encampment
on it. A little below Stockwith, where the river Idle
falls into the Trent, the county boundary line leaves the
Trent in order to include the Isle of Axholme, and goes
westwards as far as Wroot and then northwards and back
to the Trent just at its junction with the Ouse to form
the Humber, so that the rest of the river's course is within
the county. At Owston (anciently Kinnaird's) Ferry is
the site of the castle of the powerful Mowbrays, Dukes of
Norfolk. At Althorpe and Gunness the Trent is crossed by
the railway line from Grimsby to Doncaster, and another
—the New Idle—river runs into it. After passing Amcotts (of which the Amcotts family are still lords of the
manor), the hills now rise some 200 feet and come closer
to the river, and from Burton Stather (the latter word,
of Scandinavian origin meaning a landing stage) to Alkborough the Cliff is steep and well wooded. At Trent
Falls the river finally joins the Ouse, and the two form
the Humber.

There are good reasons for thinking that the course of
the river just described, from close to Newark to the
Humber, was not the original course in pre-Glacial times.
A direct prolongation of its course from Newark would
point to the great gap in the cliff at Lincoln, through
which the Witham now flows. Again, along this track
gravels are found, distinctly of river origin, and derived
from the pebble beds beyond Newark on the west. Also,
a more powerful agent than the river Witham is or ever

The Eagre running up the Trent at Gainsborough

has been, must be sought for to make such a gap, and the Trent, with its hill source and rush of many tributaries, seems the only likely river near and powerful enough to have done the work. The Trent is navigable for vessels drawing 20 feet of water as far as Gainsborough, and at certain seasons there is a tidal wave, called in Lincolnshire the Eagre (a Saxon word), which runs up some 14 miles past that town.

6. Geology.

If we turn to the geological map at the end of this volume we find that the various strata occurring in Lincolnshire run in a line roughly from north to south. This is what in geology is termed their "strike." Their "dip" or slope is towards the east, each stratum being overlaid in turn by the newer one, so that the oldest rocks are on the western side of the county. In Lincolnshire, on the surface, there are no traces of any rocks older than the Secondary or Mesozoic Group, except in the form of erratic blocks or boulders of granite brought by glacial action from Scandinavia or the mountains of the Lake district. On the west of the county a patch in the Isle of Axholme, and a long strip running north and south from near the Humber to Newark and Nottingham, represents the New Red Sandstone or Keuper marl. This layer of rock has been laid down by successive deposits in a vast inland sea which gradually became salt, like the Dead Sea in the Holy Land. Gypsum or sulphate of lime was also deposited in this bed, and it unfortunately

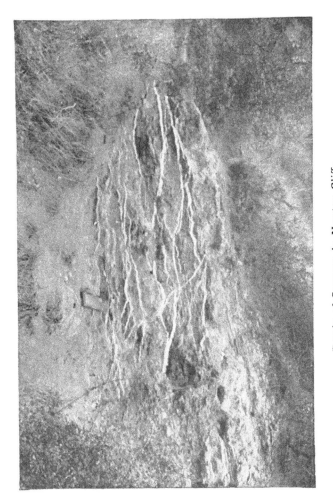

Bands of Gypsum in Newton Cliffs

was found to impregnate the water from a very deep boring (2200 feet) at Lincoln so strongly as to render it unfit for ordinary use. This boring passed through 23 feet of surface beds of soil, peat, sand, and gravel, 641 feet of lias, plentiful in fossils, 52 feet of Rhaetic beds also fossiliferous, 868 feet of Keuper marl, grey, green, red, and purple with bands of gypsum (such as are seen in Newton Cliffs in the illustration on p. 27) and 638 feet of the New Red Sandstone, into the pebble beds. The deposit of salt and brine whence the celebrated Woodhall water comes was probably produced in a similar way. This Keuper marl has not much in the way of fossils till the next layer is met—that of the Rhaetic bed (so-called from the Rhaetian Alps of Bavaria). This is full of fossils, especially remains of fish. Evidently an invasion of this inland sea by salt water caused the death of all the fish. After this seems to have occurred a sinking of the land, whereby groups of islands were formed, round which the beds of the Lias and Oolite series of rocks (called Jurassic from occurring in the Jura mountains) were deposited. The lowest layer, called the Lias, is a thick bed of shale, clay, and limestone. There is evidence of much life in the fossil remains found in this bed. From the clay pits on Lincoln hill have been taken bones of huge fish-lizards, 20 to 30 feet long, such as the Ichthyosaurus, with eyes 14 inches in diameter and a long tail; and the Plesiosaurus, with a very long neck and small head and teeth like a crocodile; and flying reptiles also. The Lower Lias in this county extends as a narrow band from the extreme south, about Sedgebrook, to the

Humber, narrowing and thinning as it passes north along the edge of the Cliff, and forming about the eastern two-thirds of the valley of the Trent. Along this route is a belt of ferruginous limestone containing hydrated peroxide of iron, which is extensively mined at Scunthorpe, where the bed is 27 feet thick. A still narrower belt on the eastern edge of the foregoing is that of the Marlstone, Middle Lias or Rockbed, which at Caythorpe (where its breadth is at its greatest) is extensively worked for iron ore. It forms a slighter escarpment west of the Cliff at Harlaxton, Gonerby, Fulbeck, and Wellborn, which dies down further north.

The Upper Lias from the valley of the Witham to the Humber is a thick bed of stiff dark blue clay, which forms the Cliff, and is much used at Lincoln for brickmaking. Ammonites and belemnites are frequent, as well as septaria, which are large nodules or spheroidal masses of calcareous marl, with crystalline divisions inside.

Capping the Cliff, and varying in breadth from 10 miles in the south of Lincolnshire to a couple of miles at the Humber, is a band of the next layer, the Inferior Oolite (a word meaning eggstone, as the rock is composed of minute grains resembling the roe of a fish). The lowest division of this is termed the Northampton Sand, and lies from 10 to 15 feet below the surface of the soil. Near Lincoln, on the east side, it is 12 feet thick, and is extensively worked for siliceous ironstone, which is sent to Scunthorpe to be mixed with the ironstone there and smelted. Above this is the well-known building stone the Lincolnshire limestone, which hardens

NAMES OF SYSTEMS		SUBDIVISIONS	CHARACTERS OF ROCKS
TERTIARY	**Recent** **Pleistocene**	Metal Age Deposits Neolithic ,, Palaeolithic ,, Glacial ,,	Superficial Deposits
	Pliocene	Cromer Series Weybourne Crag Chillesford and Norwich Crags Red and Walton Crags Coralline Crag	Sands chiefly
	Miocene	Absent from Britain	
	Eocene	Fluviomarine Beds of Hampshire Bagshot Beds London Clay Oldhaven Beds, Woolwich and Reading Thanet Sands [Groups	Clays and Sands chiefly
SECONDARY	**Cretaceous**	Chalk Upper Greensand and Gault Lower Greensand Weald Clay Hastings Sands	Chalk at top Sandstones, Mud and Clays below
	Jurassic	Purbeck Beds Portland Beds Kimmeridge Clay Corallian Beds Oxford Clay and Kellaways Rock Cornbrash Forest Marble Great Oolite with Stonesfield Slate Inferior Oolite Lias—Upper, Middle, and Lower	Shales, Sandstones and Oolitic Limestones
	Triassic	Rhaetic Keuper Marls Keuper Sandstone Upper Bunter Sandstone Bunter Pebble Beds Lower Bunter Sandstone	Red Sandstones and Marls, Gypsum and Salt
PRIMARY	**Permian**	Magnesian Limestone and Sandstone Marl Slate Lower Permian Sandstone	Red Sandstones and Magnesian Limestone
	Carboniferous	Coal Measures Millstone Grit Mountain Limestone Basal Carboniferous Rocks	Sandstones, Shales and Coals at top Sandstones in middle Limestone and Shales below
	Devonian	Upper } Mid } Devonian and Old Red Sand- Lower } stone	Red Sandstones, Shales, Slates and Lime- stones
	Silurian	Ludlow Beds Wenlock Beds Llandovery Beds	Sandstones, Shales and Thin Limestones
	Ordovician	Caradoc Beds Llandeilo Beds Arenig Beds	Shales, Slates, Sandstones and Thin Limestones
	Cambrian	Tremadoc Slates Lingula Flags Menevian Beds Harlech Grits and Llanberis Slates	Slates and Sandstones
	Pre-Cambrian	No definite classification yet made	Sandstones, Slates and Volcanic Rocks

well on exposure after quarrying. Near Ancaster and Wilsford are the chief quarries.

The Middle Oolite consists of a great thickness of clays, of which one kind is much used at Bytham in the manufacture of "clinker" bricks. Above this is the layer called the Cornbrash (so called from forming rather easily broken up soil, good for corn growing), a pasty fine-grained limestone, running in a thin line from Brigg to Stamford, with very irregular distribution in the south. Above this again is the Oxford Clay, varying between 500 and 300 feet thick, filling up the western valley between the Cliff and the Wolds, and between the Heath and the river Witham. Overlying this is the Kimmeridge Clay, which forms a large patch on the western edge of the Wold, from the valley of the Ancholme down to Dogdyke and Spilsby, edging round to within some 10 miles of Skegness, while below Lincoln it forms a belt some eight miles broad along the eastern slope of the Cliff, narrowing somewhat to its end in the valley of the river Glen.

Of the rocks forming the Cretaceous group, it may be stated at once that they were laid down in great ocean depths, for we know that the Atlantic is now doing exactly the same thing, laying down a layer of chalk (mainly composed of Foraminifera—microscopic animals and plants) on the sea floor. As time went on, and the sea added to the deposits of chalk, all Lincolnshire was buried under their weight. Later, as the land rose again, the chalk was brought to the surface and was extensively eroded by atmospheric agencies: it is now

1300 feet thick, and it had originally another 1000 feet above that.

The Wolds, a range of chalk hills, represent this formation in Lincolnshire, and stretch from the Humber, near Barton, south-east to West Keal and Burgh. They are a continuation of the Yorkshire Wolds interrupted by the Humber, and are continued into Norfolk at Hunstanton, interrupted by the Wash. A red band of chalk very noticeable in the Hunstanton cliffs is also to be seen along the line of the Wolds, from Welton-in-the-Marsh to South Grimsby, and north of Caistor.

A band of Lower Greensand, broadening out twice into a width of five or six miles, fringes the western side, and this is accompanied by a parallel though thinner band of Upper Greensand, bordered on the east by chalk some three or four times the width.

At Tealby in the middle of the Wolds are ferruginous sands 20 feet thick, devoid of fossils, and near Caistor a bed of ironstone ore has been worked since 1868. Below these are beds of clay and limestone 50 feet deep, with many fossils, especially the fine large fan-shell *Pecten cinctus*, 9 to 12 inches in diameter, the same species as that found in the Speeton Clays in Flamborough Head. These all belong to the Lower Greensand, which is also called the Neocomian, from the Latin name for Neuchatel, where there are good examples of this series.

The remaining portions of the county—the belt of marsh between the Wolds and the sea, and around Boston, the great mass of Holland—partly owe their origin and condition to glacial action. In the later period of geological history, the Glacial period or Ice Age, it is believed that England lay under a great sheet of ice much as Greenland does now, and that glaciers passing from the high lands in the north-west of Scotland and the mountains of the Lake district travelled south-eastwards across Lincolnshire. Also from Scandinavia a great glacier proceeded south and joined the Scotch and English glaciers on the east coast. Of this glacial action, the cliffs of Yorkshire and the coast of Lincolnshire are the result; for they are composed of drift or boulder clays— the debris of the glaciers which, passing over the surface of the country, wore it down, and produced a mass of mud and pebbles—with boulders of Norwegian granite and other rocks from the north-east, and of Shap granite from the Shap Fells in Westmorland from the north-west. For instance, near Barton-on-Humber has been found a large boulder of Shap granite, and another one near Ferriby in the boulder clay, where there is a small moraine. A Norwegian granite boulder was found near Thorpe Hall just outside Louth, and the "Bluestone" boulder of some four or five tons weight in Mercer Row, Louth, had probably a Scotch origin. From our county also the stream of glacier action would sometimes proceed, as near Melton in Leicestershire is a block of Lincolnshire oolite some 300 yards long by 100 yards wide. The moraines of the Norwegian glaciers still exist as a line of gravel hills from Flamborough Head into Lincolnshire, crossing the Humber at Paull.

Finally come the Fen beds, due to the action of rivers in bringing down mud and sand from the interior of the country, which are deposited along their course, especially at their outlets. These beds are composed of mud, silt, and peat, and all along the coast from the Humber to the Wash are remains of a submerged forest. Practically all Holland is typical fen. A boring near Boston pierced through 24 feet of silt, 161 feet of boulder clay, and 400 feet of Oolitic clays. The forest bed is about $2\frac{1}{2}$ feet deep, and rests on about a foot of whitish clay and sand. It crops up in several places along the coast at low water mark, and is composed of stumps and roots of oak, beech, elm, birch, holly, yew, hazel, elder, and willows. A particularly good example of this is at Trusthorpe, where it is noted that the sea made a great breach in 1777 (the forest was destroyed centuries before that), and where the original church and a great part of its parish are said to lie beneath the waves. In the fens where the forest has been uncovered, the trunks of the trees all slope to the north-east, a clear proof that in their lifetime, as at the present day, the most prevalent wind was that from the south-west.

Finally, the "blow-wells" deserve notice; they are powerful springs, rising from the chalk through the drift and alluvial deposits, and are found between Grimsby and Sutton-on-Sea.

7. Natural History.

All our existing animals, insects, and plants probably only date from after the time when Britain was over-spread with a vast and thick sheet of ice. It is very doubtful if any species can have survived the intense cold of that period. Later, when England was still joined to the continent at Dover and Calais, and across from Lincolnshire, Norfolk, and Yorkshire to Holland and Belgium, there must have been a re-colonisation of our land by animals and plants, though the latter do not depend only on land for their distribution, their seeds being often carried by birds to far distant shores. The number of kinds of animals in this island is less than that found on the continent, and in Ireland still less, showing that the channel between Britain and the continent was formed before the immigration was complete, and that Ireland being further off naturally fared worse.

Dealing with mammals first, of which there are some 50 species, there have been found in Lincolnshire antlers of the red deer[1] and bones of *Bos longifrons*, of wild horse, wolf, wild boar, and beaver. Other animals have become extinct within living memory, such as the wild cat, of which the last specimen was shot in Bullington Wood near Wragby in 1883. The pine-marten is now very rare, but weasels and stoats are common, and the polecat haunts the marshes along the east coast. The otter is much in evidence, and the badger (or brock) is more abundant in Lincolnshire than in any other midland county. Bats of three kinds are plentiful, as are hedgehogs, moles, and three kinds of shrews; squirrels abound in this county. The dormouse is recorded to inhabit the woods of south-west Lincolnshire, the harvest mouse is very rare, the field

[1] Antlers were dug up in making the lake at Hartsholme, near Lincoln, and in the peat at Barton and New Holland on the Trent.

3—2

mouse very abundant. Rabbits are exceedingly plentiful, especially on the sandy moors of the north, and hares are still very numerous and of large size on the Wolds, and on the Cliff to the south of Lincoln. Fallow deer are kept in some half-dozen parks in the county, and the herd of red deer in Grimsthorpe Park are believed to be the descendants and last remnant of the herds which inhabited the great forest of Kesteven.

Along the coast and on the sandbanks of the Wash are multitudes of the common seal, and the grey seal has been noticed also. Specimens of some eight kinds of whale have been stranded or captured along the coast, the porpoise is common, and the bottle-nosed dolphin rather less so.

With the great change in the Fens and altered conditions of the waterways, various birds have practically disappeared from the county, such as the avocets and ruffs. Grouse and blackcock, both of which latter birds used to be shot not very many years ago, are also gone. As a large arable county, it naturally breeds many partridges, mainly English, but with a sprinkling of the French or red-legged species; and, as a great sporting county, it rears very large numbers of pheasants, of which there are still some wild ones, free from imported strain. Snipe and woodcock are fairly abundant in winter, with a quantity of golden plover in the north of Lincolnshire, and the common peewit or lapwing all over it. On the salt marshes, as by Saltfleet, are generally to be found hundreds of knot, curlew, dunlin, ringed plover, redshanks, etc., with wild geese occasionally coming over.

At Scawby and Manton the black-headed gull (*Larus ridibundus*) breeds in thousands, while heronries are met with here and there, as at Evedon, and in one or two places along the Witham. The bittern has vanished, one of the last having been shot on the Holmes Common

Bittern

close to Lincoln in 1845. The barn, long-eared, and brown owls frequently breed in the county, as did the short-eared owl till some 25 years ago. The harriers (marsh and hen) ceased to breed in Lincolnshire some 50 or 60 years ago. About 1885 the buzzard ceased to nest here, though formerly common. Sparrow-hawks are quite common, and are judiciously kept under by the gamekeeper. The last kite's nest known was noted in Bullington Wood near Wragby in 1870. The hobby is a summer visitor, the merlin a winter one along the coast line. The kestrel still is common, and till 1893 a pair nested on the western towers of Lincoln Minster. The nightingale is often heard as far north as Horncastle and also, though rarely, round Lincoln.

Of the reptiles the most interesting specimen is probably the natterjack toad (*Bufo calamita*) which has been found on the sandhills near Mablethorpe.

The county is full of interest for the entomologist, though here again the drainage of the fens and an in-creased arable area have driven away many interesting kinds. Among the butterflies the chequered skipper, the brown hairstreak, the purple emperor, the greasy fritillary, and the marbled white are all notable.

At the time of Domesday Book there were about 34,000 acres of woodland, now there are some 10,000 acres more. The great forest of Kesteven was partially disafforested on its Boston side by King John in 1204, and the rest was thrown open in 1230. The submerged forests along the coast, and the remains of forests found in the peat of the fens, testify to the former covering with wood of large tracts of this county. In the parks, such as Grimsthorpe and Haverholme, are oaks, horn-beams, and hawthorns of considerable age; there is much ash on the Wolds, and Scotch firs have been planted more than 100 years ago at Fillingham and Burton. Those at

Scawby on the sandy borders of the gull-ponds give a distinctly Scotch effect to the scene. The largest willow in England is in Haverholme Park and has a girth of 25 feet at five feet from the ground. There is a large oak at Well, and another at Woodthorpe (both near Alford) whose trunk is four yards in diameter. The common oak of the county is *Quercus pedunculata*, which

Scawby Gull-Ponds

Mr E. W. Peacock has found to be the species occurring in the submerged forests and those covered in peat.

The botany of Lincolnshire, as mentioned by Dr F. A. Lees, is remarkable for presenting us with various kinds and species of plants which generally live on different soil, and at different altitudes, meeting here. This is due probably to the geographical position of the county, and to the absence of any mountainous features. The list of

Lincolnshire plants numbers nearly 2000. The sandhills along the coast still retain their covering of swordgrass and sea-buckthorn or sallowthorn (*Hippophaë rhamnoides*), the first name due to the gleaming scales beneath its leaves, known here since 1670, and remarkable for its glaucous leaves, orange berries, and sharp spikes. The salt marsh has sea-orach, and below high water level the glasswort

Moorland, near Woodhall

(*Salicornia*), and that most attractive food for wildfowl, the glasswrack (*Zostera marina*). The land round Wood-hall is still gorgeous in autumn with purple heather; the common heath, the cross-leaved kind, and ling being not at all rare. Of insectivorous plants can still be found two sorts of the sundew and the common butterwort. The rarer Alpine clubmoss (*Selaginella*), the marsh mountain ferns, the beautiful grass of Parnassus, the marsh gentian, and the bog pimpernel are all noted as inhabitants of this county. At Halton Holgate is much of the hoary cinquefoil. The woody nightshade, henbane, and dwale or deadly nightshade, are all natives of the county. The latter is mentioned by Gerarde (1636) as growing plentifully in Holland in Lincolnshire, though it is not so common now. In May the woods are carpeted with bluebells (wild hyacinth), and the lily of the valley flourishes greatly in the woods near Lincoln. The network of rivers, canals, and land drains has been choked in past times by the imported water-weed from America, the *Anacharis alsinastrum*. Lastly it may be mentioned that one species of plant, *Selinum curvifolium*, is only known to exist in this county and in the Isle of Ely.

8. Peregrination of the Coast.

As has been already stated, there are only two estuaries in or near her coast by which the rivers of Lincolnshire enter the sea—that of the Humber, which receives the Trent; and that of the Wash, which receives the Witham, Welland, and Nene. Moreover, the coast itself is very low and flat, the land near the edge of the sea being only from 4 to 20 feet above sea-level ; the low-water line is distant, and the five-fathom limit generally well out to sea. Therefore, between the Humber and the Wash there are no harbours or ports, and therefore also there are no large sea-side towns.

We may conveniently begin our peregrination of the coast at the extreme north-west corner of the county, where the Trent joins with the Yorkshire Ouse to form the Humber. This river is about three-quarters of a mile wide at the junction, and the range of hills on which Alkborough is situated comes very near to the edge of the river. This edge, passing Whitton on the way, runs east north-east for four miles to Whitton Ness, the most northerly point of the county, then makes a curve south-eastwards past Winteringham, Read's Island, and the flat lands through which the Ancholme runs, to South Ferriby, where the north-western extremity of the Wolds terminates almost on the water's edge. Probably Read's Island owes its origin to the alluvial matter brought down by the Ancholme. Close to Winteringham Haven was the north end of the Roman road, the Ermine Street. North-easterly again the coast trends, past the rather important town of Barton, which has two fine churches, and is connected by a railway with New Holland, whence there is a steam ferry to Hull, almost opposite. It is three miles from here to Skitter Ness, whence the coast runs south-easterly for 12 miles to Grimsby, about half-way being the great new dock at Immingham, where the five-fathom line approaches very close to the shore, which was one reason, no doubt, for the selection of this site for the dock.

Leaving Grimsby, which is dealt with as a fishing and general port in other chapters, an almost continuous line of houses leads to the popular watering-place of Cleethorpes, which has a large tract of sand, a long pier, bracing air, and many other attractions for the Midland excursionist, and on Bank Holidays some 100,000 people are often brought to it by train. Hence the coast runs still south-eastwards. Wider and wider get the sands exposed at low water as Tetney Haven and Grainthorpe Haven are reached, while at Donna Nook the five-fathom line has gone nearly six miles out to sea to the Sand Haile and the Rosse Spit

Skegness

Buoys. Quite possibly this may indicate land which has been overwhelmed by the sea at some past epoch, or on the other hand it may be due to the alluvial matter brought down by the Humber being deposited here. The edge of the coast from Grimsby to Skegness is composed of sand-hills, sometimes, as near Theddlethorpe, low, but a quarter of a mile wide, sometimes, as near Mablethorpe,

60 or 70 feet high but of narrow width. They are covered with sea-buckthorn and grass, and are a great nursery for rabbits. On the shore between Donna Nook and Theddlethorpe are patches of higher sand, covered with glasswort (samphire) and intersected with many channels, much frequented by various kinds of sea-birds. At the Manor House at Saltfleet, Oliver Cromwell is traditionally supposed to have slept on September 26, 1643, a few days before the battle of Winceby. On these extensive sands at low tide mirages are frequently seen.

Mablethorpe is a rising watering-place; Trusthorpe, a mile or two south, is very much smaller, but not less popular, and Sutton perhaps outrivals Mablethorpe in size and popularity. Tennyson spent much time at Mablethorpe in the earlier years of his life. "At high tide the sea comes right up to the bank with splendid menacing waves, which furnished him, five and thirty years after he had left Lincolnshire for ever, with the famous simile in *The Last Tournament*:—

> ...as the crest of some slow-arching wave,
> Heard in dead night along that table shore,
> Drops flat, and after the great waters break
> Whitening for half a league, and thin themselves,
> Far over sands marbled with moon and cloud,
> From less and less to nothing.

This accurately describes the flat Lincolnshire coast with its interminable rollers, breaking on the endless sands, than which waves the poet always said that he had never anywhere seen grander, and the clap of the wave as it fell

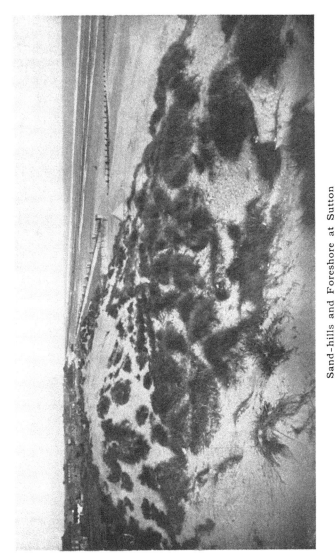

Sand-hills and Foreshore at Sutton

on the hard sand could be heard across that flat country for miles[1]."

At Sutton was the destined termination of the Lancashire, Derbyshire, and East Coast Railway, which never got farther east than Lincoln, and which has now been absorbed by the Great Central Railway. Huttoft is picturesquely situated on a piece of rising ground, and between it and the shore is the site of a proposed "Garden City," Woldsea. Chapel St Leonards is a favourite place for families to stay at, and at present it is far from the usual haunts of trippers. Ingoldmells Point is the most easterly part of the county and the five-fathom line is some five miles out to sea. At Winthorpe the Roman Bank is close to the edge of the shore. From Ingoldmells the coast-line runs south, with a slight inclination westwards, past Skegness to Gibraltar Point.

Probably a Roman fort was situated close to Skegness (part of this land was called Chesterland or Castelland in the fourteenth and fifteenth centuries) which with Brancaster on the Norfolk coast opposite would defend the Wash. Skegness owes much of its development to the Earls of Scarbrough; it is an excellent sea-side resort (out of the tripper's season), with as fine air as can possibly be found, and first-rate golf links at Seacroft. It is the resort of excursionists, very many thousands of whom are taken there annually. Just below Gibraltar Point the Wainfleet Haven or Steeping River runs into the sea,

[1] Mr Willingham Rawnsley, in *Tennyson and his Friends*, Macmillan & Co., 1911.

Wainfleet itself being a pleasant little town about two miles from the edge of the marsh.   William of Waynflete's School—reminding one of Tattershall Castle—still exists.   Here begin the great Wainfleet Sands which at low-water stretch seawards for nearly four miles, and are separated from a patch of sand called the Long Sands in the middle of the Wash by less shallow water called Boston Deeps.   From Gibraltar Point, interrupted only by the outlet of Boston Haven, the coast-line runs southwestwards to the bridge over the junction of the river Welland with the Wash at Fossdyke.   A large semicircular bend, with the convexity northwards, extends thence to the outfall by Sutton Bridge of the River Nene.   This land, and that on the east of the Nene called Wingland, has all been reclaimed from the sea. At Sutton Bridge is an unfortunately ruined dock, constructed by the Great Northern Railway.

9.   Coastal Gains and Losses.

In past times a considerable loss of land has occurred on the Lincolnshire coast.   Grimsby old harbour, it is true, has been filled up with silt, but a few miles further south part of the parish of Clee has disappeared.   Itterby has vanished, and the signal-house south of Cleethorpes has been three times set back out of reach of the sea. Saltfleet and Mablethorpe St Peter's have lost their churches and some portion of their land.   There has been encroachment also at Ingoldmells.   Leland says of " Skegnesse, a four or five miles of Wilegripe "—a port which has disappeared—"sumtym a great haven towne, the old towne clean consumed and eten by the sea." All along the coast from the Humber to the Wash can be seen at low tide the remains of a submerged forest; this is especially well seen at Trusthorpe, where the trees are of birch, fir, and oak. Along this coast the normal direction of travel of beach material is from north to south. The gains are chiefly due to the material brought to the sea by the rivers. Thus at Trent Falls in the upper Humber there were 260 grains of "warp" (i.e. silt) in the gallon, while near Grimsby there were only 30 grains per gallon at low water. In the Wash warp begins to be deposited at 5·5 feet above ordnance datum, samphire or glasswort (*Salicornia*) in another two feet, grass in a further two feet. New marsh is formed at 10·68 feet, and old high marsh at 13·15 feet, about the level of ordinary spring tide. Another valuable plant for helping the reclamation of salt marshes is the so-called "rice grass" (*Spartina stricta*), which grows below high-water mark.

Accretion is predominant in the Wash. Erosion in recent years (between 1883 and 1905) has caused a loss to Lincolnshire of 400 acres, but there has been a gain in the same time of no less than 9106 acres, partly no doubt from the material off the Yorkshire coast, but mainly from that discharged into estuaries by the rivers. In South Holland about 25,000 acres were reclaimed north of the Roman Banks by the year 1632 in the parishes reaching from Moulton to Tydd St Mary's. And in the middle of last century some 600 acres of salt " fitties " (an old Norse word for the outmarsh lying between the sea-bank and the sea) were reclaimed in the parishes of Grainthorpe and Marsh Chapel. It has been calculated that since the Norman Conquest some 330,000 acres in Lincolnshire have been reclaimed from the sea or from the waters of the fen. On the banks of the Trent, particularly in the Isle of Axholme, a special method of fertilising the land is in vogue called warping. Silt-laden tidal water is let in through sluices and drains, and allowed to stand on the land chosen for the purpose, the water being gradually drawn off with the fall of the tide. This is continued for some three or four years, and a thick layer of rich alluvial deposit is secured, making what was previously poor soil into almost the richest in England.

The protection of the coast of Lincolnshire is accomplished in various ways. Along the south shore of the Humber a substantial bank stretches between Barton and Grimsby. At Cleethorpes a sea wall and embankment more than a mile long have been erected. A low range of sand-hills with a clay foundation forms the coast protection as far as Skegness. This range has been artificially heightened and broadened with clay and faggoting, and fronted with thorn bushes in some places to accumulate the sand, as at Trusthorpe, while exposed parts are faced with a massive timber defence. Wooden groynes also have been of much service in protecting the coast between Mablethorpe and Sutton.

The coast, with its very shallow sea, is naturally a dangerous one, and the banks well out are carefully buoyed and lighted, as is the entrance to Boston Deeps from the sea, for the Wash is difficult to navigate, owing to the many sandbanks and their frequent alteration in form and size. The New Cut to Boston has fixed lights. and there are 22 lights up to the Dock entrance. Skegness Pier head has two fixed white lights. There are also light-vessels off the Dudgeon, Inner and Outer Dowsing shoals, with revolving light and foghorn, and on the last-named a submarine bell.

Off Spurnhead is a light-vessel with a revolving light, visible for 11 miles. At Spurnhead, a point of special importance to navigation, there are two lights in the light-house, the upper one visible 17 miles, flashing white for $1\frac{1}{2}$ seconds, and being obscured for $18\frac{1}{2}$ seconds. The lower light is fixed, and visible for 13 miles, showing white over the Skegness Shoal, and to the east red over the Sand Haile Buoy. This expanse of sand will be noted on the map as taking the five-fathom limit more than five miles out to sea to the Rosse Spit Buoy, east and south of the entrance to the Humber. The Humber is of course well buoyed, with a lightship on the Bull Sand (almost in the middle of the Channel), revolving white and red every 10 seconds alternately, visible 11 miles. There are also many other lights of lesser importance along our coast.

There are lifeboats stationed at Grimsby, Donna Nook, Mablethorpe, Sutton, and Skegness.

10. Climate.

The climate of a country or district is, briefly, the average weather of that country or district, and it depends upon the latitude, the temperature, the direction and strength of the winds, the rainfall, the character of the soil, the height above sea-level, and the nearness of the district to the sea.

The differences in the climates of the world depend mainly upon latitude, but a scarcely less important factor is nearness to the sea. Along any great climatic belt there will be found variations in proportion to this nearness, the extremes being " continental " climates in the centres of continents far from the oceans, and "insular" climates in small tracts surrounded by sea. Continental climates show great differences in seasonal temperatures, the winters tending to be unusually cold and the summers unusually warm, while the climate of insular tracts is characterised by equableness and also by greater dampness. Great Britain possesses, by reason of its position, a temperate insular climate, but its average annual temperature is much higher than could be expected from its latitude. The prevalent south-westerly winds cause a drift of the surface-waters of the Atlantic towards our shores, and this warm water current, which we know as the Gulf Stream, is the chief cause of the mildness of our winters.

Most of our weather comes to us from the Atlantic. It would be impossible here within the limits of a short chapter to discuss fully the causes which affect or control weather changes. It must suffice to say that the conditions are in the main either cyclonic or anticyclonic, which terms may be best explained, perhaps, by comparing the air currents to a stream of water. In a stream a chain of eddies may often be seen fringing the more steadily-moving central water. Regarding the general north-easterly moving air from the Atlantic as such a stream, a chain of eddies may be developed in a belt parallel with its general direction. This belt of eddies or cyclones, as they are termed, tends to shift its position, sometimes passing over our islands, sometimes to the north or south of them, and it is to this shifting that most of our weather changes are due. Cyclonic conditions are associated with a greater or less amount of atmospheric disturbance ; anticyclonic with calms.

The prevalent Atlantic winds largely affect our island in another way, namely in its rainfall. The air, heavily laden with moisture from its passage over the ocean, meets with elevated land-tracts directly it reaches our shores—the moorland of Devon and Cornwall, the Welsh mountains, or the fells of Cumberland and Westmorland —and, blowing up the rising land-surface, gets cooled and parts with this moisture as rain. To how great an extent this occurs is best seen by reference to the accompanying map of the annual rainfall of England, where it will at once be noticed that the heaviest fall is in the west, and that it decreases with remarkable regularity until the least fall is reached on our eastern shores.

The above causes, then, are those mainly concerned

in influencing the weather, but there are other and more local factors which often affect greatly the climate of a place, such, for example, as configuration, position, and soil. The shelter of a range of hills, a southern aspect, a sandy soil, will thus produce conditions which may differ greatly from those of a place—perhaps at no great distance—situated on a wind-swept northern slope with a cold clay soil.

The character of the climate of a country or district influences, as everyone knows, both the cultivation of the soil and the products which it yields, and thus indirectly as well as directly exercises a profound effect upon Man. The banana-nourished dweller in a tropical island who " has but to tickle the earth with a hoe for it to laugh a harvest " is of different fibre morally and physically from the inhabitant of northern climes who wins a scanty subsistence from the land at the expense of unremitting toil. These are extremes ; but even within the limits of a county, perhaps, similar if smaller differences may be noted, and the man of the plain or the valley is often distinct in type from his fellow of the hills.

Very minute records of the climate of our island are kept at numerous stations throughout the country, relating to the temperature, rainfall, force and direction of the wind, hours of sunshine, cloud conditions, and so forth, and are duly collected, tabulated, and averaged by the Meteorological Society. From these we are able to compare and contrast the climatic differences in various parts.

Speed (1627) says of this county, " The Ayre upon

the East and South part is both thicke and foggy, by reason of the Fennes and unsolute grounds, but there-withall very moderate and pleasing. Her graduation being removed from the Æquator to the degree of 53, and the Windes that are sent of her still working Seas, doe disperse those vapours from all power of hurt."

The average number of hours of bright sunshine in the year for the North Eastern Division of England from 1871–1905 was between 1100 and 1400. In 1911, a very bright year, the number of hours of bright sunshine at Skegness was no less than 1832 (the value for the district being 1597), and Rauceby had 1701. On the map recording sunshine in this year a small patch on the south bank of the Humber is marked as having 1400 hours, a third of the county 1700, and the rest up to 2000 hours.

In the annexed map, which shows the average annual rainfall, almost the whole of Lincolnshire is in the area marked " under 25 " (meaning less than 25 inches fall of rain in the course of the year). The exception is a patch along the summit of the Wolds running north-west to south-east, which comes into the higher rainfall division of "25–30." At Tealby on the Wolds the average annual rainfall, for example, was 27·35 inches, while at Lincoln only 23·34 inches of rain fell on 150 days, the average annual rainfall for Lincoln for the past 10 years being 23·27 inches. At Fulbeck on the under edge of the Cliff on 178 days there were 23·67 inches, and at Rauceby a little eastward of the last station, there were 25·66 inches. These should be compared with the average annual rainfall for Great Britain, which is 32 inches.

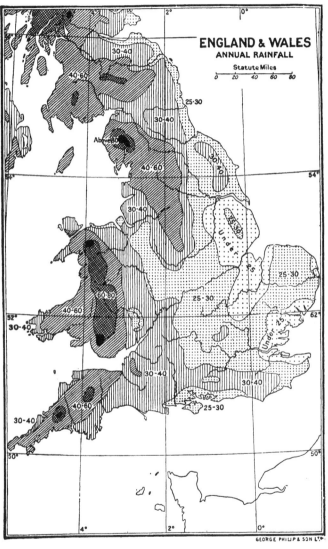

ENGLAND & WALES
ANNUAL RAINFALL

Statute Miles
0   20   40   60   80

(The figures give the approximate annual rainfall in inches.)

The number of wet days in each month varies from
11 to 17, and the wettest months are undoubtedly those
of July (2·37 to 2·60 inches), September (2·01 to 2·68
inches), August (2·57 to 3·22 inches), and October (2·47
to 3·30 inches).

The mean annual temperature for the county varies
from 47° F. at Tealby (251 feet above sea-level) to 48° F.
at Lincoln (station 58 feet above sea-level), thus being very
much the same as the average temperature for England,
i.e. 48°. The most prevalent winds are south-west, and
it is very interesting to note that the trees in the Fens
now lean towards the north-east, just as their predecessors
did hundreds of years ago before they were invaded and
swamped by peat. Owing to the fact of these winds
having swept across the Devon and Somerset moors and
the Welsh mountains and a great expanse of land before
reaching this county, they do not bring much rain ; and
the same applies to snow, which rarely falls very heavily,
except on the Wolds in a specially severe winter. The
hardest and longest frosts occur with these winds.

In the early months of the year there is a great
prevalence of easterly winds, and these, coming straight
from the North Sea, bring a large amount of moisture
with them, giving rise to grey skies, and when the wind
is south-easterly, to heavy and persistent rain. Frequently,
in the evening, a long line of cloud may be seen, lying
a little above the summits of the Wolds, showing that
some condensation is taking place from the air saturated
with moisture   And, frequently, 30 or 40 miles inland,
when an east wind is blowing, it carries a strong smell

of sea-water with it. Sea-mists are not uncommon; particularly on or after very bright hot days. And, by contrast, never is the sky more blue, or the distances more distinct, than on a day of clear bright east wind. Very few sea-side resorts of the British Isles can surpass those of the Lincolnshire coast in the splendidly bracing quality of the air.

Skating on the Fens

11. People—Race, Settlements, Dialect.

We have no written record of the history of our land carrying us beyond the Roman invasion in B.C. 55, but we know that Man inhabited it for ages before this date. The art of writing being then unknown, the people of those days could leave us no account of their lives and occupations, and hence we term these times the Prehistoric period. But other things besides books can tell a story, and there has survived from their time a vast quantity of objects (which are daily being revealed by the plough of the farmer or the spade of the antiquary), such as the weapons and domestic implements they used, the huts and tombs and monuments they built, and the bones of the animals they lived on, which enable us to get a fairly accurate idea of the life of those days.

So infinitely remote are the times in which the earliest forerunners of our race flourished, that scientists have not ventured to date either their advent or how long each division in which they have arranged them lasted. It must therefore be understood that these divisions or Ages—of which we are now going to speak—have been adopted for convenience sake rather than with any aim at accuracy.

The periods have been named from the material of which the weapons and implements were at that time fashioned—the Palaeolithic or Old Stone Age; the Neolithic or Later Stone Age; the Bronze Age; and the Iron Age. But just as we find stone axes in use at the present day among savage tribes in remote islands, so it must be remembered the weapons of one material were often in use in the next Age, or possibly even in a later one, that the Ages, in short, overlapped.

Let us now examine these periods more closely. First, the Palaeolithic or Old Stone Age. Man was now in his most primitive condition. He probably did not till the land or cultivate any kind of plant or keep any domestic animals. He lived on wild plants and roots and such wild animals as he could kill, the reindeer being then abundant in this country. He was largely a cave-dweller and probably used skins exclusively for clothing. He erected no monuments to his dead and built no huts. He could, however, shape flint implements with very great dexterity, though he had as yet not learnt either to grind or polish them. There is still some difference of opinion among authorities, but most agree that, though this may

Neolithic Implement

not have been the case in other countries, there was in our own land a vast gap of time between the people of this and the succeeding period. Palaeolithic man, who inhabited either scantily or not at all the parts north of England and made his chief home in the more southern districts, disappeared altogether from the country, which was later re-peopled by Neolithic man.

Neolithic man was in every way in a much more advanced state of civilisation than his precursor. He tilled the land, bred stock, wore garments, built huts, made rude pottery, and erected remarkable monuments. He had, nevertheless, not yet discovered the use of the metals, and his implements and weapons were still made of stone or bone, though the former were often beautifully shaped and polished.

Between the Later Stone Age and the Bronze Age there was no gap, the one merging imperceptibly into the other. The discovery of the method of smelting the ores of copper and tin, and of mixing them, was doubtless a slow affair, and the bronze weapons must have been ages in supplanting those of stone, for lack of intercommunication at that time presented enormous difficulties to the spread of knowledge. Bronze Age man, in addition to fashioning beautiful weapons and implements, made good pottery, and buried his dead in circular barrows.

In due course of time man learnt how to smelt the ores of iron, and the Age of Bronze passed slowly into the Iron Age, which brings us into the period of written history, for the Romans found the inhabitants of Britain using implements of iron.

We may now pause for a moment to consider who these people were who inhabited our land in these far-off ages. Of Palaeolithic man we can say nothing. His successors, the people of the Later Stone Age, are believed to have been largely of Iberian stock—people, that is, from south-western Europe—who brought with them their knowledge of such primitive arts and crafts as were then discovered. How long they remained in undisturbed possession of our land we do not know, but they were later conquered or driven westward by a very different race of Celtic origin—the Goidels or Gaels, a tall, light-haired people, workers in bronze, whose descendants and language are to be found to-day in many parts of Scotland, Ireland, and the Isle of Man. Another Celtic people poured into the country about the fourth century B.C.—the Brythons or Britons, who in turn dispossessed the Gael, at all events so far as England and Wales are concerned. The Brythons were the first users of iron in our country.

The Romans, who first reached our shores in B.C. 55, held the land till about A.D. 410; but in spite of the length of their domination they do not seem to have left much mark on the people. After their departure, treading close on their heels, came the Saxons, Jutes, and Angles. But with these, and with the incursions of the Danes and Irish, we have left the uncertain region of the Prehistoric Age for the surer ground of History.

Of the Celtic population of this county at the time of the Roman invasion but few traces are left, thus contrasting greatly with what has happened in counties such as Somerset, Cornwall and the wilder parts of Wales, and the Lake district, where the Brythons (hence the name Britain) fled before the Roman advance and later from the Saxons. These Celts, belonging to the tribe of Coritani, have left little impression on the names of places (Lincoln itself being an exception), and probably none on the actual people of Lincolnshire. On the other hand the Saxon invasion and settlement must have been complete early in the sixth century. With respect to the Danish invasions in the ninth and tenth centuries the case was otherwise, and the lion and the lamb would lie down at length peaceably, as after all they were essentially of the same racial stock. This can be seen by frequent intermingling of names. The Danish settlements and their advance in Lincolnshire may be traced in four special directions on the map by means of the Scandinavian names of towns or villages. No less than 195 of these names end in *by* (originally = a single dwelling-house), while 76 end in *thorpe*, which represented a collection of houses, or village. One advance was certainly made from near Grimsby westwards and southwards, and another from the Trent eastwards, and the two streams would meet somewhere about Caistor. Again, on the coast from Saltfleetby to Skegness the names of Scandinavian origin are thickly spread, and so on to the Wolds around Spilsby and Alford and Horncastle. Moreover, the stream of invaders and settlers must have come up the Fossdyke from Gainsborough to Lincoln.

There is no great distinction nowadays to be found between the two races of Saxons and Danes in Lincolnshire. In a list of citizens at Lincoln in the fourteenth century, Old Norse and Saxon names are fairly equally represented. The country folk are generally speaking fair-haired, and, like David, ruddy of countenance. The ordinary language in the county is much the same as on the east coast and the south of Scotland, and is Saxon, added to and modified, but not supplanted, by Norse or Danish. There is a distinct difference between the dialects of the parts of the county divided by the Witham,

between north Lincolnshire, which was the home of
the Lindiswaras, and south Lincolnshire, where the
Gyrwas dwelt. Of the northern dialect (which approxi-
mates fairly closely to that of its neighbour Yorkshire)
Tennyson's *Northern Farmer* and other poems in dialect
will serve as excellent examples. Probably many of the
local pronunciations of words are original and right, in
reality. For instance, road (where the vowels do not
make a diphthong) is pronounced as a dissyllable, ro-ad,
instead of as in our modern parlance " rode," leaving out
the *a* altogether. One exception possibly from what
has been said above as to the disappearance of all Celtic
traces in the county, may be found in the way in which
the shepherds, or at all events some of them, number
their flock. This notation, pethera, pimp, dik, bumpit, yan
a bumpit (4, 5, 10, 15, 16), is strikingly like that in the
Celtic system, and that in use in modern Welsh, the
Welsh equivalents being pedwar, pump, deg, pymtheg,
unarbymtheg. Both systems start again at 15, and do
not go further than 20, when a " score " was cut on the
tally, and the counting commenced over again[1].

There are still some enduring traces of Huguenots in
the county. From both France and Flanders there was
a stream of immigration into England after the massacre
on St Bartholomew's day, 1572, and more especially after
the Revocation of the Edict of Nantes in 1685. The
Flemish, being trained to drainage working, helped
Vermuyden in draining the Fens. In 1626, for the

[1] An almost identical notation occurs in parts of Cumberland and
Westmorland—Ed.

use of those foreigners working on Hatfield Chase, a
chapel was built at Sandtoft near Belton, in the Isle
of Axholme, wherein services were held alternately in
French and Dutch. Thence, in the troublous times of
the Great Civil War, many of these settlers were driven
away, and made for Thorney. One family has been
traced from Hatfield Chase to Thorney, whence it
spread to Fleet, Crowland, Brothertoft, Swineshead, and
Sutterton, and probably many others are in the same
category.

12. Agriculture—Cultivations, Stock.

Lincolnshire is pre-eminently an agricultural county,
and a great proportion of its large area is devoted to the
production of various crops. In 1911 out of its total
of 1,705,293 acres, arable land accounted for no less
than 1,003,743 acres, while permanent pasture occupied
517,925 acres. The county possesses a larger acreage of
barley than any other in the kingdom, and produces rather
more than two bushels per acre above the average. Her
wheat acreage also is the largest in England, yielding nearly
six bushels per acre above the average. For acreage of
oats she ranks between Yorkshire and Devon, with eight
bushels per acre over the average. Her acreage of peas
is the largest in England, that of beans between Suffolk
and Essex, with over seven bushels to the acre over the
average. The Lincolnshire potato-growing area is
20,000 acres larger than that of Yorkshire, the next on
the list ; and she ranks third among the counties for

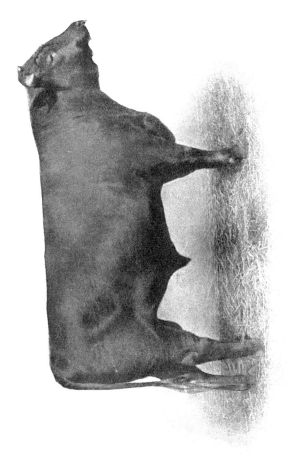

Lincolnshire Red Shorthorn Bull

turnips and swedes. It is interesting to note that in the ten years 1895–1904 the Lincoln corn-market was first in the list for oats three times, and for barley once, Norwich being generally the first corn-market in the kingdom, with London and Peterborough coming next. In 1910 Lincoln ranked sixth on the list for wheat, second for barley (about 100,000 quarters behind Norwich, and about the same in front of Berwick), and sixth for oats, Stamford being third. About three thousand acres are allotted to orchards, and about two thousand acres to small fruit, especially strawberries, currants, and goose-berries. There are nearly five thousand small holdings from one to five acres each, and about ten thousand between five and fifty acres. Along the roads leading from Lincoln to Branston, Navenby, and Low Brace-bridge, can still be seen small one-storied houses which have each had about a rood of ground attached to them. These were early precursors of "Three acres and a cow," and were built by Fergus O'Connor, the Chartist.

Lincolnshire has a great reputation for breeding and raising stock, and takes third place (after Yorkshire and Devon) among the counties of England in the number of cattle she possesses. A special breed (constituting about 90 per cent. of the cattle bred in the county)—the Lin-colnshire Red Shorthorn—is becoming well-known; it is of a well-defined type, with much wealth and evenness of flesh, and with great milking qualities. As regards sheep her position is the same, ranking after Yorkshire and Northumberland. The typical Lincolnshire sheep is the largest and heaviest of its kind in the kingdom, has a good growth of bright fleece, is of great hardiness of constitution, and is much used for crossing with other varieties, such as the Leicester. It is also in great request for exportation to Argentina, Australia, and

Lincolnshire Ram

New Zealand, where it is crossed with the Merino. Consequently immense prices have been paid—as much as 1000 guineas—for Lincolnshire rams. There are large sheep-fairs held annually at Lincoln and Corby, but the numbers are very much smaller than in years gone by. At Sleaford wool-fair in 1911, no less than 15,000 fleeces were for sale. For pigs the county takes fourth place, the majority being of the Large

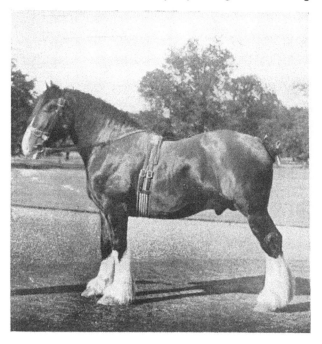

Lincolnshire Shire Horse

White breed, which attain a very great size, 50 stone being a not infrequent weight; but a native kind, the Lincolnshire curly-coated, is rapidly advancing into favour.

As becomes a county wherein two famous horse-fairs, Lincoln and Horncastle, are held, Lincolnshire has a great repute for breeding horses for riding, driving, and for heavy work; she ranks second to Yorkshire for the number of her horses. The shire horse has been a Lincolnshire production for generations past. It is interesting to note—as snowing how customs change—that in 1566 it was observed that in Lincolnshire there were "few draught horses; the carriage of that county standeth most by oxen." Hence comes one of Shakespeare's two allusions to the county in Mr Justice Shallow's enquiry " How a good yoke of bullocks at Stamford Fair?"[1]

Of special cultivations, there is now nothing very particular to chronicle. As we shall see in the next chapter, in old days, before the drainage of the fens, very many geese of excellent quality were bred there. On the Heath were very large warrens of silver-grey rabbits (some still exist near Santon, close to Frodingham) whose skins were very marketable. Some 300 or 400 acres of fen were also devoted to the produce of cranberries. Mustard is extensively grown for seed about Holbeach and Spalding, and the Isle of Axholme is well known for its vegetables, such as celery. Fields may also be seen of white poppies for " poppy-heads " and for the production of opium. Flax also is cultivated round Epworth and Crowle. A favourite Lincolnshire vegetable, Good King Henry or mercury, is extensively

[1] The other being "as melancholy as......the drone of a Lincolnshire bagpipe," both from the play of *Henry IV*. Pepys was entertained in 1667 to drink and the bagpipes by Sir Freshville Holles, a Lincolnshire M.P.

grown and used as a rather coarse spinach, and the glasswort of the coasts (*Salicornia*) is used for pickling. Around Boston some woad (*Isatis tinctoria*) is to be seen, and there were two woad-growers registered in the *Lincolnshire Directory* for 1909. The blue dye is obtained from the root-leaves, which are crushed in a mill by rude conical crushers dragged round by horses, and the

The Woad Industry : Balls drying in the Sheds

pulp thus made is worked up into balls and laid out for some weeks to dry. These are then thrown in a heap in the dark, mixed with water, and fermented, being left for a considerable time before being packed into casks for sale. This dye is now always used with indigo.

Near Boston also, in the last few years, flower farms, producing narcissus and tulips, have come into vogue.

In the Isle of Axholme, round Haxey in particular, and on the higher levels, the land is cut up into parallel strips called selions, about a rod wide and half an acre in extent. As these belong frequently to different owners (one man for instance owned 40 acres in 100 different plots in one village), they are diversified in crops.

A Field of Tulips at Spalding

On the Heath or Cliff, which extends nearly from the Humber to south of Grantham, the soil is thin and near the oolite rock. The fields are large, often walled in, and the older farm buildings generally of stone. The rotation of crops is carried out with marrowfat peas, wheat, roots and barley, which is of the best quality. Carrots are much grown, especially where the soil is sandy.

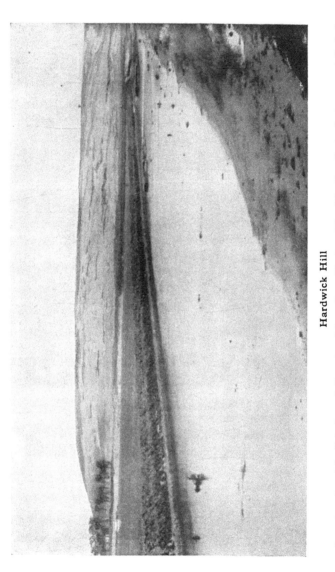

Hardwick Hill

A continuation of the Gainsborough Escarpment from Ferry Flash—with a Warping Drain, in progress, at the foot

On the Wolds also the soil is very thin, hardly more than a foot deep above the chalk. The farms are large, running from 300 to 1500 acres, the largest of all being at Withcall. A four-course rotation of crops is strictly carried out, wheat (or oats rather commonly of late years), turnips, barley (with seeds sown in it), and wheat, or oats again.

Much of the north Marsh is under permanent grass, and is some of the most valuable grazing ground in the kingdom. Further south a good deal of the land is now tilled, and produces mustard, potatoes, and corn. The growth of straw and yield of grain are very large indeed.

The Fens are now so well drained that they are rather short of water in a dry summer. Most of the land is under arable culture, the corn crops being mainly wheat or oats, occasionally barley, or beans or peas. Turnips, mangolds, potatoes, and carrots are the chief root crops.

13. Industries and Manufactures.

In the centuries immediately succeeding the Norman Conquest the chief industry of Lincolnshire was the preparation of wool, as far as regards the dwellers on the Cliff, Heath, and Wold. It was no doubt due to the export of wool that the port of Lincoln ranked fourth among the ports of the kingdom in the sixth year of King John (1204), while Boston ranked second, Grimsby tenth, Barton eleventh, and Immingham twelfth. In 1291 Lincoln was made a "staple" town, wherein the wool was sold, weighed, and certified, and then sent down the river Witham to Boston. In 1361, however, the latter port had the "staple" transferred to it, to its great advantage, and to the great discontent of Lincoln, whose inhabitants vainly petitioned to have it restored. Nearly a hundred years before, Boston had stood at the head of the Customs returns for several years (1278–1290). In spite of the complaints of Lincoln's decaying trade, it must have been fairly prosperous, since it is estimated to have had a population of 5000 in 1377, when it was the sixth largest town in the kingdom, and in 1503 it was by assessment actually fourth in the list. It had had a guild of weavers for centuries, and in the middle of the thirteenth century Lincoln was celebrated for its manufacture of scarlet cloth. Later, the colour always associated with Lincoln was green, as is mentioned in Spenser's *Faery Queene*, Drayton's *Polyolbion*, and the ballads of Robin Hood, but few clothiers seem to have been in the city in the earlier years of the sixteenth century. The weaving trade probably went abroad during the stormy times of King John and Henry III, and never came back to Lincolnshire again, though the actual wool export trade must have lasted on well into the times of Queen Elizabeth, as the fine manor house at Bassingthorpe testifies, being built by a merchant of the staple, Thomas Coney, who had 1000 sheep in 1569.

Later efforts to stimulate the working-up of the wool in the county do not seem to have met with much success.

Several attempts were made at Lincoln both to establish
knitting and spinning schools, and also clothworking in
the sixteenth and seventeenth centuries, but they all
came to nothing. In 1561 Thomas Trollope proposed to
Cecil, Queen Elizabeth's powerful minister, to set up
mills at Stamford for the beating of hemp and the
manufacture of linen and canvas cloth, but they met
the same fate as those dealing with wool. At Belton
House there is still some excellent tapestry made at
Stamford in the eighteenth century. In 1787 an Annual
County Ball was established for the encouragement of
native woollen manufactures, for the first two years being
held at Alford, but ever since at Lincoln. From its
origin it is still often called the "Stuff Ball." The ladies
used to wear stuff gowns and the gentlemen stuff coats,
waistcoats, and breeches, according to the late Sir Charles
Anderson, who adds that stuff was little worn after 1820.
From the custom of the Lady Patroness choosing the
colour or colours of the ball, it is also sometimes called
the "Lincoln Colour Ball."

In the Fens, before they were drained, one great
source of income was from the very abundant supply of
wild fowl. For these decoys were used, of which there is
still one left at Ashby near Burringham on the Trent.
In 35 seasons at this decoy the total catch was nearly
100,000 wild fowl, of which nearly half was contributed
by mallard and teal[1]. A great number of wild fowl
are still taken in nets along the sand-flats on the Wash.

[1] *Victoria County History of Lincolnshire*, vol. ii, article Wildfowling, from
which all that follows on this subject is taken.

Both Fuller and Camden wrote enthusiastically of the
wealth of wild fowl in Lincolnshire, and Pennant in 1768
refers to this county as "the great magazine of wild fowl
in this kingdom." Another source of profit to the "Fen
Slodgers," as the men were called, was from the reeds and
rushes, which were gathered for thatching before tiles

Wildfowling in the Fens

and slates came into ordinary use. Camden says that
a well-harvested stack of reeds was worth from £200 to
£300. Dr Johnson was told that a roof thatched with
Lincolnshire reeds would last 70 years. It is easy to
understand why the fenmen were so keenly and so
tenaciously determined to resist the drainage of the
Fens.

Cutting Reeds for Thatching

It is interesting to note that, in 1585, an order was made for "xij or xvj Stiltmen in the countie of Lincolne furnished with either of them two paire of the highest of Stiltes and the longest poles that are or maie be used with the said stiltes to be sent over into the Low Contryes to the Erle of Leicester."

Geese, as already stated, were in past years kept in enormous numbers in the Fens, both for their feathers, which provided the penman's quill before Birmingham had popularised the substitute of steel, and for stuffing beds, as well as for food. There are still many kept in the districts around Spalding.

As would be expected in an agricultural county, there are large crushing-mills for linseed and for other kinds of cake for feeding stock. These are at Lincoln and Gainsborough, and there are manure works at Lincoln and Saxilby. Flour-mills driven by steam are fairly common in most of the large towns, and have superseded windmills, which are disappearing fast, and are not replaced. Forty years ago there was a row of seven or eight along the cliff between Lincoln and Burton, now two only are left. In a barley-producing county malting is naturally a very prominent industry, and the enormous new malt-kilns and houses at Sleaford are probably far the largest in the country.

In Saxon times, certainly after the days of King Edwy, there was a mint in Lincoln which struck coins there of all the succeeding kings up to the Norman Conquest. When Domesday Book was compiled the Lincoln mint paid £75 to the king, a larger sum than was paid by any other mint in the country.   After the Conquest, coins struck at Lincoln are known of all the kings (Kings Richard I and John being excepted) down to King Edward I.

This city also was one of the provincial assay towns, as was ordained by a statute of the year 1423 (the second year of King Henry VI) that each of the towns mentioned should have divers " touches," i.e. marks, and further that no goldsmith should work silver of worse alloy than the sterling, and should put his mark upon it before he " set to sell," under the same penalties as those obtaining in London.   No mark is known peculiar to Lincoln.   But there is a mark—a capital I on a capital M in a florid type of shield, almost invariably alone—which has been found on fifty communion cups which, except in one instance, are all in this county.   It is almost certainly a Lincoln maker's private mark.   These cups have paten covers, and on the paten foot is frequently inscribed the date, i.e. 1569 in 19 instances, 1570 in two, and 1571 in one case.   Other cups unmarked, but of the same date and style, may safely be attributed to the same maker, whose name, unfortunately, is not known.

Towards the end of the first half of the nineteenth century agricultural-implement works were started in Lincoln.   From small beginnings these works have grown and prospered exceedingly, their buildings and shops covering many acres and giving employment to some thousands of workmen.   They produce portable and fixed engines, boilers, traction engines, road rollers, pumps, threshing-machines, hay and straw elevators, maizeshellers and chaff-cutters, oil engines, gas engines, steam navvies, engines for mining and electrical purposes, and steam wagons. In some of these works all the machinery is electrically driven, and in all the best and latest developments in tools, workshops, etc., are found. Plough-works, malleable-iron works, and wire-works also produce a large

Iron Works, Lincoln

output, and employ many men. At Grantham also are large iron-works making much the same class of machinery, with a particular leaning to oil and gas engines. At Gainsborough are large foundries and iron-works, with like products in agricultural and other machinery. The smelting industry, which has attained such large dimensions at Scunthorpe, will be noticed in the next chapter.

Iron Works, Gainsborough

6

At Brant Broughton much excellent work has been done recently in wrought iron, artistically handled.

At both Gainsborough and Lincoln are large wood works, and, as will be seen later, wood holds an important position in the import trade of the Lincolnshire ports. A comparatively new industry which employs many hands, chiefly women and girls, in Boston, Lincoln, and other towns of this county, is pea-picking, in which the peas are sorted out in sizes and qualities and packed for sale in boxes. Another large industry is that of the feather factories, wherein the feathers supplied by farmers and poultry dealers are sorted by machinery and then purified by steam, the residue going to form a valuable manure.

Of places in the county that have given name to any product there are only four—Lincoln, as mentioned above, Grantham (for gingerbreads), Boston, and Torksey. Boston seems to have been known in Elizabethan times for its drinking-vessels, as Bishop Hall in his Satires refers to the " palish oat frothing in *Boston* clay[1]."

At Torksey[2] a china manufactory was established in 1803 by William Billingsley, with his son-in-law, George Walker. Billingsley had previously been many years at Derby working for Duesbury, the proprietor of the old Derby works. The business at Torksey only lasted five years. Billingsley was an admirable flower and landscape painter on china.

---

[1] *Victoria County History of Lincolnshire*, vol. II, article Industries, p. 388.

[2] *Associated Architect. Societies' Report*, vol. XXIII, pp. 153—156, Dr O'Neill.

In the early days of lawn tennis, a considerable industry was established in Horncastle for the manufacture of the racquets used in the game. But this has died out in the last 20 years.

14. Mines and Minerals.

The first place among Lincolnshire mines must be given to Scunthorpe and Frodingham in the north of the county. The fact that the ironstone there was sufficiently rich to make it worth smelting was only realised about the year 1855, when the late Lord St Oswald (then Mr Rowland Winn) first opened quarries. The ore was at that time taken to the river Trent, and shipped to iron-works in Yorkshire. The first blast furnace in the district was erected about 1864, and others followed shortly afterwards. There are now five firms who smelt iron on the spot, and in addition to the ore used by them, a very large quantity is sent to iron-works in Yorkshire and Derbyshire. The area in which ironstone is being dug extends from Ashby on the south to Thealby on the north, a distance of about seven miles—its widest part measuring about a mile and a half—and it includes portions of the parishes of Ashby, Brumby, Frodingham, Scunthorpe, Flixborough, Normanby, and Burton. The ore is a fossil-bearing limestone in the Lower Lias and contains the iron in the form of hydrated peroxide. The bed, where it attains its full thickness, is about 30 feet deep ; it has a slight north-easterly dip, and the quarries are all situated on its outcrop, so that the available thickness diminishes from east to west, according to the degree of denudation to which it has been subject. Towards the east the bed dips under the scarp of the hill; but it was reached in shafts and borings near Appleby station at a depth of 300 feet, still being nearly 30 feet thick. The stone is all got in open quarries. It is covered with blown

Frodingham Iron and Steel Works: Scunthorpe

sand varying in depth from a few inches to about 30 feet, containing in places beds of peat. This is removed by digging and burrowing, or in some cases by mechanical means. The ironstone is got by drilling and blasting. The percentage of metallic iron varies in the different bands that make up the full thickness of the bed; some of the richest yield upwards of 30 per cent. and some are too poor to treat. It has been calculated that on an average two tons of coal produce one ton of metal. The stone contains in itself sufficient lime to act as a flux, and a siliceous component is furnished by the ironstone of the Northampton Sands, quarried at Greetwell by Lincoln.

The ore is smelted in large blast furnaces, and the result is mostly disposed of in the form of pig iron. But, a few years ago, one firm at Frodingham built steel works and rolling mills, using the Siemens-Martin method. The Lincolnshire steel is of very high quality, suitable for rolling into thin sheets or drawing into wire. Some extensive new works wherein both iron and steel will be made are nearing completion at Flixborough.

At Caythorpe are considerable open workings of the Middle Lias (Marlstone) ironstone.

At Greetwell and Monks Abbey, just east of Lincoln, as already mentioned, is quarried the siliceous ironstone found in the lowest layer of the Lower Oolite, known as the Northampton Sands. The ironstone is worked partly in the open when there is little soil above, but chiefly by galleries driven into the Cliff, with narrow-gauge rails and trucks, on which horse-traction is being superseded by small locomotives. The ore is reddish brown at the outcrop and gets bluer in colour the deeper the tunnel goes in. The yield of metal is from 28 to 40 per cent. The soil is replaced in the open workings, and has been covered with allotments, etc.; in the other workings the galleries have fallen in and produced a very irregular surface.

Near Claxby and Nettleton the Middle Neocomian layer of ironstone, about six feet six inches thick, has been worked by galleries driven into the side of the hill. The workings began in 1868. The ore is almost entirely made up of small and beautifully polished oolitic grains of hydrated peroxide of iron. It is a calcareous ore, yielding from 28 to 33 per cent. of metallic iron, and is useful for mixing with the clayey ores of the coal-measures. From the presence of slag with charcoal and bits of pottery it is evident that this bed of ore was known to and worked by the Romans during their possession of this country.

The county is not rich in other minerals, coal not yet having been tapped to any practical result. The chief building stone is the Lincolnshire Limestone, an oolitic rock worked near Ancaster and Wilsford, at Haydor, and near Grantham, of which many churches and houses in Kesteven are built, including Lincoln Minster. It hardens on exposure and forms a most excellent building stone. Many churches on the Wolds and in the Marsh are built of the beautiful local grey-green sandstone (of the Lower Neocomian series), which unfortunately is rather perish-able. In some instances the white chalk is used for building, as at Legbourne Church, where the smooth white surface suggests at a distance unglazed white tiles. The clay on Lincoln hill and below the Cliff is extensively used for brick-making, and at Little Bytham are works for making so-called "clinker" bricks, which are specially hard and used as fire-bricks.

15. Fisheries and Fishing Stations.

There are several different methods adopted for
catching sea-fish. Among the most important is that
of trawling, in which a triangular net with a bag is
towed along the sea-bottom. This, therefore, can only
be done when the sea-floor is fairly smooth and sandy
and without rocks. By this method the bottom-feeding
fish are caught. Steam trawlers are gradually superseding
the sailing vessels. In winter the east coast trawlers—
with which we are chiefly concerned—fish the Dogger
Bank, which lies in the North Sea about midway between
the coast of Yorkshire, Durham, and Northumberland, and
that of Denmark, and each vessel takes its own catch into
port, having a well on board wherein the fish can be kept
alive, while in harbour special boxes are placed in the
water for the same purpose. In the months of February
and March cod are very plentiful along the coast north
of the Humber, and from 150 to 200 boxes of haddock
are often landed at Grimsby by a steam trawler after a
week's fishing on the Dogger Bank. In summer they
fish along the Danish, German, and Dutch coasts in
fleets, wherefrom a steamer takes the catch to port.
The fish caught are chiefly flat-fish, halibut, turbot, brill,
soles, and plaice, and those possessing the quaint names of
witches and megrims, with some cod, haddock, whiting,
hake, gurnard, and red mullet. Of late years, trawlers
have gone farther a-sea for their quarry ; to Iceland,
where are large plaice and haddock, and south to Vigo
Bay, in the Bay of Biscay, where hake are plentiful.

The second kind of sea-fishing is by lines and hooks, and is carried on over much the same area as the former, especially over the Dogger Bank and Cromer Knoll, but fish are caught by this method at any depth, so that the state of the sea-floor, whether rocky or sandy, is of no consequence. Mussels and whelks are extensively used for bait, the lines are some eight miles long with 4580 hooks on each, and are shot across the tide. Cod and haddock are the main catch. The Faroe, Shetland, and Iceland fishing-grounds are worked by large steamers from Grimsby, which bring back enormous numbers of halibut, with ling, cod, coal-fish, and skate. Fishing by head-lines, with generally two hooks, is familiar to every visitor to our coast, and is practised near the shore.

The third kind of fishing is by drift-nets, which are hung suspended vertically across the tide, and through the meshes of which the fish (herrings on the north-eastern coasts, pilchards round Cornwall) get their heads, but cannot get them back, owing to the gills.

Grimsby, from a small town with 9000 inhabitants in 1860, has sprung into the foremost position in Europe and probably the world, as a fishing port. She had in the 1911 census a population of 74,663, a magnificent fleet of 564 steam fishing vessels (with tonnage 41,648), which is more than one-third of the entire number of vessels in the United Kingdom. The fish-market is two miles long, and in 1910 there were landed at Grimsby no less than 3,491,000 cwts. of wet fish, valued at £2,528,000, as well as £6000 worth of crabs, oysters, and other shellfish. In the preservation of all this fish much ice is necessary, and hence there is a very large ice factory in the town. An effort is being made (1911) to make this port also one of the most important in the kingdom for curing and pickling herrings, as is done at Yarmouth and elsewhere.

One of the earliest notices of Boston as a fishing port occurs in the year 1325, when orders were sent to buy and provide for the King's use in the markets of St Botolph,

Fish-pontoon, Grimsby

ten thousands of stockfish and styfish. In 1907 there were at Boston 95 fishing vessels, which employed 433 men and boys, and in 1910, 88,075 cwts. of wet fish, valued at £66,242, and £7835 worth of shellfish were landed at this port.

Oysters used to be very abundant near Saltfleet, and they are still numerous near Cleethorpes, where they grow to a large size. Under Boston jurisdiction in the

Wash are some six to eight square miles of mussel "scalps."
A good deal of shrimping is done along the Lincolnshire
coast, and smelts were caught in large numbers, as
also were grey mullet, near Wainfleet Haven, but the
fishermen complain that these have decreased in late years,
and the fishing is now chiefly for flounders and dabs.
One enemy of fish, the seal, has recognised this part of
the coast as an excellent feeding place, and although several
have been killed and captured, the fishermen have had
to leave this locality to the seals. In 1912 there was a
colony of 500 seals in the Wash, doing great damage to
the fisheries and the Eastern Sea Fisheries Association
offered 5s. for each seal killed. These, of course, are of
little commercial value either for skin or oil. In the north
of the county the gate-posts at the entrance to a farm
are occasionally formed by huge whale jaws, testifying to
the prevalence of whale-fishing years ago, before the
whales were nearly exterminated.

16. Shipping and Trade.

The principal port in Lincolnshire, as has been in-
dicated in the preceding chapter, is that of Great Grimsby.
Situated on the south bank of the Humber, not far from
its entrance into the North Sea, Grimsby is exceedingly
well placed for the promotion of river, coast, and foreign
trade. Its jurisdiction extends from Skitter Ness in
Goxhill parish, on the north-west almost opposite Hull,
to Trusthorpe drain on the south, where that of Boston

begins.   The port has undergone several vicissitudes.
In the reign of King Edward III, it supplied 11 ships
and 170 seamen for the siege of Calais.   With the
gradual silting-up of the harbour its shipping trade
declined (in 1588, probably there was not a ship above
100 tons at Grimsby), until the latter end of the eighteenth
century, when steps were taken to improve the harbour.

Grimsby Docks

A dock of about 14 acres was finished in 1800, when
the population (in the following year) was only 1524 in
number.   With the advent of the Manchester, Sheffield,
and Lincolnshire Railway—now the Great Central Railway—on the scene in 1848, the new era of commercial
prosperity opened for Grimsby.   The Royal Dock, of
25 acres, had its first stone laid by the Prince Consort
in 1849, and received its name when Queen Victoria

visited the town in 1854. There is also the Alexandra Dock (48 acres, partly including the old dock), a uniting dock between them, an old fish dock of 13½ acres, and a new one of 9½ acres. The dock gates and locks are moved by hydraulic power, derived from a stately tower 335 feet high. A new dock at Immingham, six miles higher up the river, where the five-fathom line comes

Immingham Dock, Grimsby
(*Showing Coal Hoists*)

very near to the shore, is rapidly approaching completion, and will be one of the largest, if not the largest, on the east coast. It is a deep-water dock, in a land-locked harbour, with a deep-water channel, and consists of a square basin 1100 feet square, with two arms 1250 feet long and 375 feet wide, making a total water area of 55½ acres.

Grimsby has also ship-building and engineering works and a very large trade altogether apart from fish and their belongings. In 1910, she exported 1,611,220 tons of coal, her steam fishing fleet shipped 833,420 tons, and other vessels 270,025 tons, making a grand total of coal passed through the port of 2,714,665 tons. In the same year she imported 305,478 loads of timber and wood goods valued at £776,857, butter to the value of £3,124,154, corn, grain, etc. £120,114, eggs £311,878, and bacon £422,597. In the same year the total value of her imports (£12,615,959) and her exports (£18,958,924) was no less than £31,574,883.

Boston attained to a much greater position in medieval times than Grimsby, owing doubtless to its being the port at the entrance of the Witham into the Wash, whereby a very large trade in wool was carried on. In 1204, of the tax on the fifteenth part of land and goods of the merchants at this port, Boston paid £780, coming second in the kingdom to London with £836. In 1279 the Hanseatic merchants were trading here, and merchants from Ypres, Cologne, Caen, and Ostend had houses in the town. The reputation of its great fair was widespread. The canons of Bridlington Priory, for example, regularly attended this fair to buy their wine, groceries, clothes, etc., as did those of Fountains Abbey. In 1336, a grant of protection was given to a number of German merchants and 14 ships to attend the fair. When King John of France was confined in Somerton Castle, he procured spices from Boston, and rented a cellar of wine from William Spaign (the name still exists in Spain Lane) in the town. In 1359, to King Edward III's fleet of 710 ships, with 14,151 men, Boston contributed 17 ships and 361 mariners. In 1361 the staple of wool was removed to Boston, and no doubt contributed very largely to the commercial growth and prosperity of the port. In 1575, the authorities received praise for having captured some pirates, who were handed over to be dealt with by Lord Clinton, vice-admiral of the court. The port seems nearly to have been brought to ruin in Elizabethan days, probably by the silting up of the Witham and the shifting sands of the Wash. The dissolution of the monasteries also, and some quarrels between the Esterlings and the townsfolk, helped in the decay of commerce. In 1751, owing possibly to the diversion of some fen drains from the Witham, a small sloop of 40 to 50 tons and drawing six feet of water, could only sail to and from the town at spring tides.

Early in the nineteenth century things were better, coals came by the Witham, grain was sent to London, and there was some trade with the Baltic. In 1812, Boston had 177 vessels, one of 412 tons, and in that year and the preceding one, one-third of the oats received in London was shipped from Boston. A new dock, 825 feet long by 450 feet wide, was made in 1882, a new channel was cut from Lynn Well to Boston Deeps, the bed of the Witham was deepened in 1896, and a new channel made to deep water. In dealing with a sluggish river like the Witham, there is always a difficulty with regard to locks and the proper scouring of the outfall. Without locks it is almost impossible to keep sufficient

The Docks, Boston

water in the river for the necessary traffic, while the
presence of a lock at the outfall means a considerable
silting up of the channel on the seaward side of the
lock, and an inefficient scouring of the channel. The
jurisdiction of the Port of Boston extends from Trusthorpe drain on the north to Fleet Haven outfall or
Sutton Corner in the south-east. Her trade consists of
imports of timber from the Baltic, and of grain, cotton,
and linseed from the Mediterranean, the Black Sea, and
America. There is a regular line of steamers to Hamburg
twice a week.

In 1909, 369 steam and sailing vessels of 177,630
tons entered the port, 383 vessels of 176,062 tons cleared,
and 53 sail and steam vessels belonged to the port with a
tonnage of 3684. The Deep Sea Trawling and Steam
Trawling Companies do a large trade in fish. Boston's
exports of coal, coke, and other fuel were 180,415 tons.
Her imports amounted to the value of £891,126, con-
sisting of barley £88,343, maize £93,655, refined sugar
£360,627, and timber £98,175.

Little need be said of the former ports of this county.
Lincoln was among the most considerable ports in the
time of King John, and its decline was no doubt due
chiefly to the removal of the staple therefrom to Boston,
and to the state of the river Witham. Huttoft apparently had some trade once, now there is practically no
estuary[1]. A list of Lincolnshire ports in 1342 gives
Lincoln, Boston, Saltney, Saltfleetby, Wainfleet, Barton-

[1] Leland says "at Huttoft marsch cum shippes yn from divers places
and discharge."

on-Humber (whence there used to be a ferry to Hull,
before New Holland was made the ferry station by
the M. S. & L. railway), Grimsby, Burton-on-Stather,
Whitton, South Ferriby, Stroyten, North Coates,
Swynhumber, Tetney, Wrangle, Surfleet, Spalding,
Torksey, Gainsborough, and Kinnard's Ferry. Of these
Torksey, situated at the junction of the Fossdyke with
the Trent, has been of considerable importance both as
a stronghold and a port in Roman and medieval times.
Gainsborough still does a considerable trade by water.
Spalding is included in the Boston trade. Skegness,
according to Leland, in the time of King Henry VIII,
had been "at sumtyme a great haven towne...a Haven
and a Towne waulled, having also a Castelle."

17. History of the County.

The first notice of this part of our country is that of
Claudius Ptolemaeus about the year 120 A.D., in which
he mentions the British tribe of Coritavi or Coritani.
This tribe inhabited the site of the existing counties of
Lincoln, Rutland, Leicester, part of Nottingham, Warwick, and Derby, and had as its chief towns, Lindum
(Lincoln), and Ratae (Leicester). Under the Roman
domination Lincoln was first a fortress, and then a colony.
Several tombstones have been found there to soldiers of
the Ninth or Spanish Legion, and Lincoln can still show
a Roman City gate (Newport Arch), part of her Roman
walls and the ditch surrounding them, and other remains.

The Roman roads and canals are mentioned in other chapters, and the Roman villas, discovered at positions far from any military protection, give an idea of the peace which must have prevailed under the Roman rule. The Roman banks which protected the land from the sea are still to be seen. These banks and the necessary dykes

Newport Arch, Lincoln

were neglected when the last legion departed in 426, and once more large tracts of Fen were left to be covered by floods.

To resist the invading Picts and Scots, King Vortigern is said to have invoked the aid of Saxons, Jutes, and Angles (the last-named settled in Mercia, of which Lincolnshire was a part). Meeting the northern invaders at Stamford

he defeated them with much slaughter, being assisted by
the Saxon forces under the command of Hengist and
Horsa.   At Caistor, King Vortigern is supposed to have
met Hengist's daughter, Rowena, who afterwards married
him.   She is related to have poisoned her stepson
Vortimer, who died in 475 and was buried at Lincoln.
In this year Hengist ravaged the country, and captured
London, Lincoln, and Winchester, which were regained
by the British under Ambrosius in 487.

With the Saxon invasion Paganism replaced Christianity.   Two hundred years later, the Venerable Bede
recounts the re-introduction of Christianity.   In the
year 628 he says "Paulinus also preached the Word to
the Province of Lindsey...and he first converted the
Governor of the City of Lincoln, whose name was
Blecca, with his whole family."   In 678, Egfrid, King
of Northumbria, captured Lindsey from Wulfhere, King
of Mercia, and had Eadhed ordained the first Bishop of
Lindsey.   In Doomsday Book are named over 200
churches as in existence in Lincolnshire at the time of
the Norman Conquest, which testifies to the widespread
establishment of Christianity in these parts.

But another Pagan invasion, that of the Danes, was
to sweep over the country towards the end of the eighth
century, and Lincolnshire was particularly exposed to
their attack, from the facility with which they could land
on the coast, or sail up the Humber to Gainsborough and
Torksey, and thence along the Fossdyke to Lincoln.
Direct evidence of these incursions and settlements is
furnished by the names of places such as Mable*thorpe*,

Trust*thorpe*, and Saltfleet*by* along the coast, and Saxil*by*, Kettle*thorpe* and Skelling*thorpe* along the Fossdyke. Lincoln also was one of the five towns which were under the Danelagh (Leicester, Nottingham, Derby, and Stamford were the others, to which Chester and York were added later). This grouping of towns replaced the kingdom of Mercia, and Lincoln seems to have represented the Lindiswaras (dwellers in Lindsey on the higher land) as Stamford did the Gyrwas (who lived in the Fens). Each town was ruled by its own Earl with his separate host, twelve lawmen administered Danish law, and a Common Court of Justice existed for the whole confederacy.

In 869 the army of the Pagans (i.e. of the Danes) under Hubba and Hingvar having made some stay at York, at the close of the winter passed over by ship into Lindsey, and landing at Humberstone (possibly *Hubba stone*) ravaged the whole country. Ingulph, whose authority is of no great weight, describes a battle fought on St Maurice's day (Sept. 22) 870, at Stow Green near Threekingham, in which the Danes were beaten with great slaughter. At Threekingham, near the church, there is still a large mound in which some of the slain were buried, and a piece of land in the parish is called Danes Field or Danes Hill to this day. In 873, when the Danish forces wintered at Torksey, peace was made with them. In 911 Mercia was infested with Danes, and in 937 occurred the battle of Brunanburgh, one of the greatest in the long struggle between Saxons and Danes. Possibly Burnham near Thornton Curtis in

North Lincolnshire was the site of this battle, wherein King Athelstan with his brother Edmund Atheling gained a very decisive victory over Constantine King of the Scots, with whom were many Danes under Anlaf.   The Danish King Sweyn died at Gainsborough in 1014.

Entrenchment at Burnham

In 1068 William the Conqueror visited Lincoln and ordered the erection of the castle.   He distributed lordships freely among his followers, this county being divided up between twenty-three Normans, of whom Gilbert de

Gant, Odo Bishop of Bayeux, and Alan Earl of Richmond, took the lion's share. In the next year Earls Edwin and Morcar, with two hundred and forty ships, landed on behalf of Edgar Atheling on the Lincolnshire side of the Humber and were almost all captured by a strong force of the king's friends from Lincoln. The Empress Maud, coming to England in 1140 and asserting her claim to the Crown against King Stephen, took up her residence at Lincoln, which was well provisioned and fortified. The city (and probably the castle) was soon besieged and taken by King Stephen, but the Empress had managed to escape previously. Next year Ranulf Earl of Chester and his half-brother William de Roumare, whom King Stephen had created Earl of Lincoln, captured the castle by an ingenious trick, and were besieged by Stephen, who regained the city. Earl Ranulf, escaping, brought back Robert Earl of Gloucester and a large army to raise the siege. A battle ensued on the north-western slopes of the city, and partly through treachery ended in the complete defeat of Stephen, who had fought most gallantly. From the comparative ease with which this battle was won, it got the name of "The Joust of Lincoln." Stephen, having been exchanged for the Earl of Gloucester, again besieged the castle, of which he obtained possession in 1146. Later, it was once more attacked unsuccessfully by Earl Ranulf. The Empress Maud's son, King Henry, was crowned again in 1158, in Lincoln, but in a suburb without the walls. Here, too, in November 1200, King John met William, King of Scotland, who swore fealty and did homage to him. On November 23, the body of St Hugh, Bishop of Lincoln, was received at Lincoln by King John and three archbishops and thirteen bishops, and buried in the Cathedral on November 26. In 1216 came the closing

Jews' House, Lincoln

scene of John's restless and evil life, when he left Kings Lynn with a powerful army and lost all his baggage in crossing the river Nene, a part of the Wash[1], he himself

---

[1] This was close to Sutton Bridge, a large tract of land having been reclaimed from the sea since that date.

with the army only just escaping. October 13 he was at Swineshead Abbey, the next day he proceeded to Sleaford Castle, and thence to Newark Castle, where he died on October 18. A long siege of Lincoln Castle had been carried on by the Barons who were on the side of the French Prince Louis. This was relieved on May 19, 1217, by an army under the command of William, Earl Marshal (attended by the Papal Legate), who threw Fulk de Bréauté with crossbowmen into the castle, and forced open the west gate of the city. After much hand-to-hand fighting, the party of the Barons and the French was decisively beaten, their leader the Comte de Perche slain, and the city and close given up to plunder. Owing to the great amount of booty gained the battle was nick-named "Lincoln Fair."

In 1255 the Jews of Lincoln were accused of having crucified a Christian boy called Hugh, and King Henry III and his Queen were at Lincoln to investigate the case. In 1265 the first writs of general summons to Parliament were issued, and Lincoln, London, and York were the only cities expressly named to send up two burgesses. On October 6, 1280, the beautiful Angel Choir of the Minster received the body of St Hugh, translated there with much state ceremony. On December 2, 1290, King Edward I was in Lincoln for the burial of the viscera of his dearly loved Queen Eleanor, who had died at Harby. The first of the Eleanor Crosses stood just outside the city south gate. An important Parliament was held in Lincoln in 1301, which dealt with the pretensions of the Pope to dispose of the

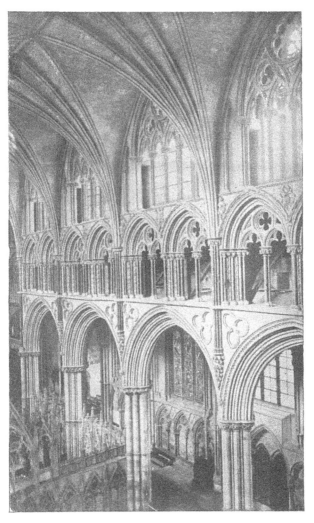

The Angel Choir, Lincoln Minster

kingdom of Scotland. Parliaments were also held at Lincoln in 1304, two in 1316, and one in 1327, and it has been thought that the oak seat of state now in the Chapter House at Lincoln was made for the king's use at one of these Parliaments. A Parliament was summoned to meet at Stamford in 1309, where also Councils were held in 1326, 1337, and 1392.

Education was not neglected in medieval Lincolnshire, for it appears that there were at least eleven schools in the county existing in the first half of the fourteenth century, besides the grammar schools at Lincoln, which date from 1090. And several of Lincolnshire's sons, whether by birth or adoption, had great influence in, or were great benefactors to, both our ancient universities of Oxford and Cambridge. Robert Grosseteste (Bishop of Lincoln 1235–1253), was one of the foremost teachers of his time, a great scholastic and ecclesiastical reformer, and Chancellor of Oxford ; Richard Fleming (Bishop of Lincoln 1420–1431), founded Lincoln College, Oxford, in 1427, which was refounded by Archbishop Rotherham (Bishop of Lincoln 1471–1480), who was also Chancellor of Cambridge, and a great benefactor to the library of that University and to King's and St Catharine's colleges; William Alnwick (Bishop of Lincoln 1436–1449), was another great benefactor to King's College. In 1457 William of Waynflete, Bishop of Winchester, founded Magdalen College, Oxford ; he also built and endowed a school at his birthplace, Wainfleet. In 1512, William Smith (Bishop of Lincoln 1496–1514), rebuilt and practically refounded Brasenose College, Oxford ; and in 1516,

Richard Fox, Bishop of Winchester, who was born at Ropsley near Grantham, founded Corpus Christi College, Oxford. At one time it was not beyond the bounds of possibility that Lincolnshire might have rejoiced in a University of its own, as in 1333 there was a large secession of masters and students from Oxford to Stamford, and this attempt was not ended till two years later, after the aid of the Queen and the Bishop of Lincoln had been obtained, and after three royal monitions and the seizure of the seceders' goods.

With the Wars of the Roses Lincolnshire was not, fortunately, much concerned. In 1470, however, a Lincolnshire man, Sir Robert Wells, eldest son of Richard Lord Wells and Willoughby, was persuaded by the King-maker, the Earl of Warwick, to raise a large force of men for the Lancastrian cause. Apparently he drove Lord Burgh out of his house at Gainsborough and burnt it, and with some 30,000 men proclaimed King Henry. But King Edward IV managed to get Lord Wells and his son-in-law Sir Thomas Dymoke into his power, and set out for Stamford with them and a strong force. He also made Lord Wells order his son to desist from his undertaking, but as this order was set at nought, he proceeded to behead Lord Wells and Sir Thomas Dymoke, and marched against the Lancastrian army. So savage were the commanders of this latter force (Sir Robert Wells and Sir Thomas de la Launde) with King Edward's action that they would not wait for the arrival of the Earl of Warwick, but commenced the battle at Hornefield near Empingham on March 12. Over 10,000 men were slain, the Lancastrians were defeated and both their leaders were taken prisoners.

In 1536 occurred the Lincolnshire portion of the rebellion against King Henry VIII, called the Pilgrimage of Grace. There were several causes for this outbreak, the suppression of the smaller monastic houses being among the most powerful. On October 1, Nicholas Melton, shoemaker (known consequently as Captain Cobbler), and others, took possession of Louth Church, so as to stop the jewels of the church (as they said) being given up to the king. In the course of a few days risings of the same description took place at Horncastle, Caistor, and elsewhere in Lindsey. By October 6 some 25,000 men were encamped round or in Lincoln, and a letter was sent to the king. His answer was read to some 300 of the troops in the Chapter House where the gentlemen were collected and where they were nearly massacred, only managing to escape out of the south door of the Chapter House. After some discussion, and a diplomatic address from a Herald, the forces dispersed. About twenty persons suffered for this rising, Moigne (Recorder of Lincoln), the Abbot of Kirkstead, the Abbot of Barlings, and eleven more, were tried by a commission in March 1537, and hanged or gibbeted in various towns of the county, Lord Hussey being executed at Lincoln.

In the great Civil War between King Charles I and his Parliament, Lincolnshire held an important position midway between the Royalists of Yorkshire and Nottinghamshire, and the Puritans of East Anglia, and on her heath or wold Cromwell found a training-ground for his Ironsides. One of the first overt acts was that of Lord Willoughby of Parham (one of a younger branch of the Willoughby d'Eresby family), in calling out the militia in June, 1642. This was followed by a brisk correspondence with the king, who spent two nights in Lincoln and was received with much loyalty. It was perhaps on this occasion that he presented the third sword to the city. He once more passed through Lincoln on August 20 on his way to Nottingham, where he raised his standard two days later. In 1643 Lord Willoughby was made by Parliament Sergeant-Major-General for the county. In April of that year Crowland stood a siege of about a fortnight, Oliver Cromwell being one of the capturing commanders. Next year, having been recaptured in the interval, it stood a siege by the Parliamentarians of two months' duration.

One of the greatest thorns in the side of the Parliament in this district was Newark, which was successfully held for the king till he surrendered to the Scottish army. He spent his last day as a free man at Barn Hill House, Stamford, before going on to Southwell, May 3, 1644. In February 1643, a combined force from Derbyshire, Nottinghamshire, and Lincolnshire, attacked Newark, but were repulsed, owing, it was said, to the half-hearted conduct of the Lincolnshire commander. As a retort, Colonel Cavendish, a young and brilliant cavalry leader, captured Grantham on March 22, and took Stamford, defeating Cromwell. Again Cavendish was victorious at the battle of Ancaster, and in a third skirmish the Royalists also won, but in a fourth, on the road from Grantham to Newark, Cromwell defeated Cavendish. As a result of this, Grantham and Lincoln must have become Parliamentarian, as the latter place had its castle and walls put into a state of defence by order of Parliament. An abortive attempt to seize Lincoln for the king, promoted by the Hothams of Hull, took place on Sunday, July 2. Gainsborough, with its commander the Earl of Kingston, was surprised on July 20 by Lord Willoughby, and the Earl, being sent as a prisoner down the Trent to Hull, was killed by a cannon ball fired at the pinnace by the Royalists on the banks. On July 28 Cromwell, fresh from the capture of Stamford and Burghley House, met Cavendish at Lea, near Gainsborough, where the latter was defeated and killed. Then Cromwell had to retire before the advance of the whole army of the Earl of Newcastle, and Lord Willoughby surrendered Gainsborough, left Lincoln (as he considered the fortifications were too slight to be any protection) and retired to Boston. But by the commencement of October, Fairfax's cavalry had been transported from Hull into Lincolnshire, and were joined by troops under Lord Willoughby, Cromwell, and the Earl of Manchester. On October 11, at Winceby, five miles south-east of Horncastle, these forces met the Royalist troops, who intended to raise the siege of Bolingbroke Castle close by, and decisively routed them. "Slash Lane," between Winceby and the high road, still commemorates the slaughter which took place. On October 24 Lord Manchester captured Lincoln, but next year Prince

Rupert's brilliantly successful attack, March 22, 1644, on the forces besieging Newark, led to the evacuation of Lincoln and Sleaford, and to the dismantling of the defences of Gainsborough.

But the whirligig of time soon brought its revenge, and on May 3, 1644, the Parliamentary forces under Lord Manchester attacked the lower part of the City of Lincoln and carried it, and after waiting a day, as there was a great rainfall which made the slopes very slippery, captured the castle and upper town by storm on May 5 with surprisingly little loss.

In 1648, orders were sent to put Tattershall Castle and Belvoir Castle into defensible condition, as these were the only two places capable of defence in or near this county. This was done to protect them against raids from Pontefract Castle, which was in the hands of the Royalists. In Lincoln the only stronghold was the Bishop's Palace, which on June 30 was attacked, captured, and burnt by the Royalists, under Sir Philip Monckton. This force left Lincoln and a few days later was followed by Colonel Rossiter with a powerful detachment, who gained a complete victory over them at Willoughby, near Nottingham.

Since the Civil War there has been little in the way of history to record in connection with our county.

18. Antiquities—Prehistoric, Roman, Saxon.

In Lincolnshire there have been found, as yet, no traces of the earliest races of mankind on this island, and the stone axes, knives, spear-heads, arrow-heads, and the like that we do find all belong to the Neolithic, or New Stone Age. The people of this date are known as dolicho-cephalic (i.e. long-headed) and were buried in long mounds or barrows, of which there are examples at Swinhope and elsewhere in the county. Five boats, in each case made out of a single tree-trunk, have been dug up, two at Lincoln, one at Scotter, and two at Castlethorpe near Brigg. Two very early boats or canoes were discovered in Nocton parish in 1811, but it is difficult to assign a date to these objects. Around Scunthorpe, on the peat moors, have been found quantities of small, beautifully-made flint implements.

Of Bronze Age man there are many relics, such as swords, celts, spear-heads, daggers, shields, and pottery of good design. The people of this age were brachy-cephalic (i.e. round-headed) and were interred, generally after being burnt, in barrows of a round shape, of which examples in Lincolnshire are numerous.

Many of the more important earthworks in the county probably belong to this age. Such is the camp above Honington, which has a triple rampart and two ditches; the circular encampment at Ingoldsby; the huge earthworks called the Moats at Irby-on-Humber; the camp at Kyme, with a double rampart; the circular

Bronze Implements

(Found at Caythorpe in 1884)

mound at Kingerby (wherein three British skeletons were found) enclosed by a ditch with a square embankment outside it; the Castle Hills, Gainsborough, afterwards used by the Danes if not due entirely to them; the great earthworks at Withern; and the great mound at Hoe Hill, near Fulletby. At Tetford Lock are some hut-circles.

British Camp at Honington

Of the Iron Age, traces are found in the pre-Roman smelting works at Manton, and in many swords, spearheads, and shields, found in the Witham. A beautiful Romano-British shield also found in that river should be mentioned.

Of the Roman occupation of Lincolnshire there are many remains. Lincoln, once a Roman colony, as already stated, retains portions of its Roman ditch, rampart, and wall, the only existing Roman city gate, Newport Arch, and the lower part of a long colonnade in Bailgate. Here also a Roman milestone of the date of Victorinus was found, and hypocausts or heating apparatus, altars, tombstones (chiefly to soldiers of the Ninth, the Spanish Legion), tesselated pavements, etc. Other Roman stations in the county were at Horncastle (Banovallum on the river Bane), where the Roman ditch and part of the wall are evident, also at Caistor and Ancaster, with the ditch well shown, where an altar for incense, a milestone, a group of Deae Matres, and Romano-British graves have been found. South Ormsby was a watch or outpost camp, between Burgh and Caistor; Yarborough camp was near Melton Ross, and Alkborough overlooked the junction of the Trent, Ouse, and Humber. Many tesselated pavements have been discovered, as at Roxby, Scawby, Winterton, and Horkstow, in the north of the county, Scampton (a few miles north of Lincoln), and Little Ponton, near Grantham. These, belonging to private houses, show the state of security of the country during the Romano-British period.

The chief Roman roads, which are mainly in use now, are the Ermine Street, which beginning at Pevensey enters Lincolnshire at Stamford, skirts Grantham, and passes through Ancaster to Lincoln, where it joins the Fosseway. From Lincoln the Ermine Street runs almost due north to the Humber, and is in full use for the first 17 miles. Four miles north of Lincoln a branch road, called Till Bridge Lane, leaves it on the west and runs to the ford at Littleborough on Trent (Agelocum or Segelocum). The Fosseway extends from the south of Devon to Newark, and enters our county just beyond Brough. It forms the county boundary here for about a mile. Another Roman road enters the county at West Deeping and runs fairly straight to Sleaford, while another connects Lincoln with Horncastle.

The Fossdyke

Evidence of Roman engineering also remains in the Fossdyke, a canal joining the Trent at Torksey with the River Witham at Lincoln; and the Cardyke, which beginning near Peterborough, runs northwards, skirting the junction of the higher ground and the fen, to Washingborough, three miles from Lincoln, where it joins the Witham. The Roman banks, placed so as to resist the encroachments of the sea in the south-east of the county, have already been mentioned.

Of both Saxon and Danish antiquities there is not much to record, save in the way of churches, parts of which, especially in towns, mark the work of the former people. Both races have left their mark more on the speech of the people and the names of places. The fortifications called the Mainwarings, with a double fosse, near Swineshead, have been attributed to the Danes, and no doubt they occupied several of the earthworks, such as the Castle Hills, Gainsborough, already mentioned. The existence of coins minted in Lincoln in the reigns of King Alfred and (missing out the reigns of Edward the Elder, Athelstan, Edmund, Edred, and Edwy) all the succeeding monarchs to the Norman Conquest, is a proof of the importance of the city in Saxon and Danish times.

19. Architecture—(*a*) Ecclesiastical.

The material used for building churches in Lincoln-shire was almost invariably stone. Near Lincoln and southwards on the Cliff this was the local oolite lime-stone, of which the most famous quarries were at Ancaster and Wilsford ; further south and south-west much of the stone came from Barnack in Northampton-shire ; on the Wolds and in the Marsh, a green sandstone was in use, not very durable, but weathering delightfully from an artistic point of view. In some cases, as at Legbourne, the white chalk itself is used, giving, as already mentioned, a curious effect as of unglazed white tiles.

The first churches of which we have any remains in this county are of Saxon building. Their builders had but an imperfect knowledge of construction in stone, and imitated rudely the Roman buildings which existed in England. A very early plan of a Saxon church had a central tower, with a chancel on the east and possibly a baptistery on the west side, as at Barton (St Peter's), Broughton, and Hough on the Hill. Later, the usual plan was a square and tall western tower, with a midwall shaft in the belfry windows (of these there are over 30 instances in this county) and a tiny chancel, opening by a very narrow arch into the nave. The walls were rather thin and roughly built, the corners of the nave had " long and short work " (i.e. large stones set alternately vertically and horizontally as in the nave of St Mary le Wigford in Lincoln, and those of the parish churches of Bracebridge, Cranwell, and Ropsley), there were no buttresses, but occasionally the wall or tower was ornamented with strip panelling, as in the lower part of the tower at St Peter's, Barton, and in the western tower arch of Stow.

With the Norman Conquest came in a new style from the continent, the Romanesque or "Norman." It was a massive style, with very thick walls, round-headed arches for doorways, windows, and arcades, sturdy pillars, with large capitals, semi-circular vaulting, flat roofs, and square towers with low pyramidal tops.

The middle portion of the west front of the Minster at Lincoln is a good example of Norman work, while the beautiful doorways, highly ornamented, and the lower

Lincoln Minster: West front

half of the two western towers, are good specimens of later work of this period. The nave and chancel (later) of Stow are Norman, Clee church has an early Norman arcade on the north side of the nave, and a later one on the south side; at Whaplode the chancel arch and eastern bays of the nave are good Norman work, the three western ones being Transitional (between Norman and the next period).

For about 70 years, from 1180 to 1250, a further development of architecture took place. It was characterised by much thinner walls, high-pitched roofs, pointed arches and vaulting, which instead of having the weight supported by thick walls, has it spread scientifically over large buttresses on the aisle walls, and flying buttresses to the clerestory walls. This is the first period of "Gothic" called First Pointed or Early English. Other features are piers of grouped slender pillars, often of marble (at Lincoln Minster much Purbeck marble was used), conventional foliage round the capitals, long narrow lancet-headed windows, and rich deep mouldings round doors and windows. Of this period St Hugh's Choir and the great transepts in Lincoln Minster are early examples, built before 1200, when St Hugh died. The nave is also a superb work of about 30 years later, and in lightness of design and elegance of proportion it is very hard to equal. Kirkstead Chapel, Bottesford (near Brigg), Grimsby parish church, and St Mary's Weston are almost entirely Early English, as is the beautiful west front of Crowland Abbey, which resembles the west front of Wells Cathedral. In this period the towers were first

Parish Church, Grantham

made to carry spires, as at Frampton, Rauceby, and
Sleaford, and the tower and spire (of timber, lead-covered)
of Long Sutton is an admirable specimen.

After the middle of the thirteenth century the
Decorated Period began and windows became broader,
divided up by bars of stone called mullions, with their
heads ornamented by patterns of tracery. The gradual
introduction of this can be well seen in the grouping of
two window openings under one arch, when the space
between all three arches is perforated, as in the triforium
of St Hugh's Choir at Lincoln. The most perfect and
sumptuous example of this Early Decorated or Geometrical
Gothic is the Angel Choir of Lincoln Minster (see p. 105).
The north aisle of Grantham Church is also of this period,
and probably owes much to the Angel Choir at Lincoln.
By 1300 was finished one of the great architectural
glories of the county, the tower and spire of Grantham,
281 feet high. Not many years afterwards, the Broad
and Rood Tower of Lincoln Minster, with its timber
and lead spire (rising altogether to a height of 525 feet—
excepting Old St Paul's, quite the loftiest spire in Europe
at that time) was completed.

Several of the finest churches in the county, especially
round Sleaford, belong to the later Decorated Period, and
of these perhaps Heckington is the typical queen. The
window tracery is more elaborate, a favourite pattern
very prevalent in this county being reticulated or like
network; the mouldings are rich, the vaultings more
intricate, and pinnacles and spires are adorned with
crockets and finials of well-wrought foliage. The foliage,

of internal work, often closely resembles natural leaves, such as vine, oak, and sometimes holly, as can be seen in the South Choir aisle of Lincoln Minster, and on the shrine of Little St Hugh.   The grand parish church

Boston Church

of Boston is chiefly of this date, while Ewerby (with a fine broach spire), Helpringham, Silk Willoughby, Croft, Welbourn, and Winthorpe may serve as admirable and diversified examples.

In 1349 came the widespread destruction of life by

the plague called the Black Death, and church and all other building was stopped for half a century or more.

The next period of Architecture, the Perpendicular, is entirely English and is not found abroad. It is characterised by much flattened arches, elaborate vaulting (the so-called fan-vaulting is not infrequent in flat roofs), and the prevalence of vertical lines in the tracery of windows and panelled ornament, which has given the period its name. Several grand churches in the Marsh were built at this time, as Grimoldby, Marsh Chapel, Theddlethorpe All Saints, Tattershall Collegiate Church, and Sedgebrook. Towers without spires now became frequent, the superb "Stump" (as it is locally named, from being spireless) of Boston, 293 feet high, Great Ponton, of a kind more frequent in Somerset, and several in marsh-land being instances. Claypole, Donington, Leadenham, and Stamford All Saints have good Perpendicular spires, while Louth has a tower and spire 300 feet high, only second to that of Grantham. This spire cost £305. 7s. 6d. It was begun in 1501 and finished 14 years later, the weather-cock being made out of a copper basin taken two years previously from the Scottish king by the men of Lincoln at Flodden Field.

Not much church-building took place for many years after the Reformation, but of the so-called "Classical" architecture there is the admirable Minster library and colonnade in the north side of the cloisters, built by Sir Christopher Wren in 1674. The parish church at Gainsborough (except the tower) was rebuilt in 1745; St Peter-at-Arches, Lincoln, a typical "City" church,

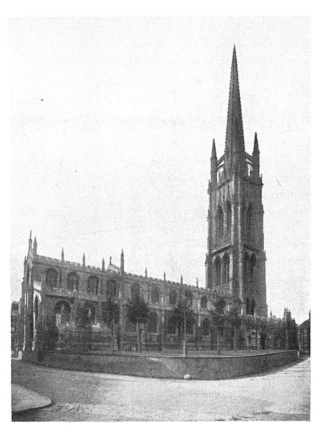

Louth Church

in 1724, by Abraham Hayward; and the diminutive church at Cherry Willingham about 1770.

Towards the middle of the last century began the Gothic revival; very many of the churches of the county have been restored, and a considerable number rebuilt, and among the finest of the new ones may be mentioned those of Nocton and Fulney, St Swithin's Lincoln, Morton, and Revesby.

There is evidence of the former existence of no less than 124 Religious Houses, Monasteries, Priories, Friaries, and Hospitals in Lincolnshire. At the time of the re-introduction of Christianity into this county in the Saxon era, monasteries were established as outposts to assist the missionary work. Only two of these, out of several in this county, were re-established after the Danish invasion and the Norman conquest. The Benedictine Abbey of Bardney, 10 miles from Lincoln, founded by King Ethelred and Queen Osthryd of Mercia in 697, was rebuilt in Norman times, and its ground plan, with bases of pillars of excellent Norman and Early English work, and many beautiful memorial slabs, has recently been laid bare. Crowland Abbey, also Benedictine, was founded in honour of St Guthlac about the year 714, by Ethelbald, King of Mercia[1]. The splendid west front, dating from 1171, the north aisle of the nave (used as the parish church), a Perpendicular tower, with part of the nave, a grand Norman arch over the western

---

[1] After its destruction by the Danes it is interesting to note a gift to the Abbey from King Canute of 12 polar bear skins for the altars, to keep the priests' feet warm.

Crowland Abbey

crossing, and a stone screen underneath it, remain. It was far the wealthiest house in the county, being worth £1093. 15s. 10½d. a year at the time of its suppression by King Henry VIII. In 1114 was founded another Benedictine Priory at Frieston. Many of these houses were founded by the great noblemen or landowners, and endowed with the tithes and presentations of rectories in the county and elsewhere. When the religious houses were dissolved by King Henry VIII and their income, lands, and buildings were given to his favourite courtiers, a small part only being devoted to education, the parish often retained the part of the monastic church wherein it had been wont to worship, as in the case of the north aisle at Crowland and the nave at Frieston.

At Stamford there still remain the west front and five arches of good Norman work of the nave of the church of St Leonard's Priory. After Crowland the most remarkable remains of monastic buildings are the ruins of the fine Perpendicular gatehouse of Thornton Abbey, which belonged to the Austin canons and was founded in 1139. Both Crowland and Thornton were presided over by mitred abbots, who consequently had seats in the House of Lords. For a brief space after the dissolution of the Abbey a college existed at Thornton, founded by King Henry VIII. The parish church at South Kyme preserves the south aisle of another house of Austin canons, while the parish church of Bourne is the nave of the church of Bourne Abbey, which belonged to a reformed branch of the Austin canons.

The parish church of Sempringham is the north

Thornton Abbey

aisle and part of the nave of the Abbey founded by St Gilbert
of Sempringham in 1139. This order, the Gilbertine,
was the only order founded in England and consisted of
Augustinian monks and Cistercian nuns, with lay brothers
and sisters, kept strictly apart though living under the
same roof. In their churches a wall ran from east to
west completely dividing the monks and lay brothers
from the nuns and lay sisters. There were 10 houses of
this order in the county. The Knight Templars had
five preceptories in Lincolnshire. Of these a fine tower
at Temple Bruer is alone left. After their downfall their
property passed to the Knights Hospitallers, or Knights
of St John, who had three other houses as well.

The church of the Grey Friars at Lincoln, built in the
thirteenth century, still exists, with a later vaulted undercroft inserted, and after having been used as a grammar
school for some centuries, has now been taken over by
the Corporation as a museum. A small portion of the
hospital of St Giles at Lincoln is also existing. The
picturesque building of Cantilupe College, founded in
1367 for the warden and seven chaplains to commemorate
the souls of the founder and his wife in Lincoln Minster,
is just south of the great south door. Of Tattershall
College, founded by Ralph Lord Cromwell, only the
splendid Perpendicular church is left.

The beautiful fifteenth century churchyard cross at
Somersby ought to be mentioned, and the fine series of
sepulchral monuments to the St Paul family at Snarford,
the Willoughbys at Spilsby and Edenham, the Monsons
at South Carlton, and the Heneages at Hainton. At the

east end of the Minster there are interesting monuments to the Burghersh and Wymbush families, and a reproduction of Queen Eleanor's original altar tomb and effigy.

20. Architecture—(*b*) Military.

In various parts of the county there are remains of entrenched camps, which may date from British, from Roman, from Saxon, or even from prehistoric times. But the building of castles began practically with the reign of William the Conqueror, who ordered the erection of a great number throughout the land and began that at Lincoln after his visit in 1068. It may therefore be taken as a fair example of the plan and arrangements of a castle in the late eleventh and early part of the twelfth centuries. It occupies most of the area of the south-west quarter of the first Roman city, on the top of the hill, where most probably the original British city of Lindum stood, and is roughly quadrangular, the south and west walls being on the lines of the Roman walls. It is guarded by a broad and deep dry ditch (for it is situated really on the limestone rock), and by a massive bank of earth, 50 to 80 yards broad and from 20 to 30 feet high, sloping steeply externally. On the middle line of these mounds are strong walls, 8 to 10 feet thick and from 30 to 40 feet high, which date probably from before the time of King Stephen. Originally, no doubt, the mounds were topped by a palisade. The main entrance to the

Lincoln: the Castle Gateway

castle is by the eastern gate, which is Norman, but has had a later (Edwardian) arch and towers affixed to it, and had a little outwork or barbican, now destroyed. On the north-east angle is a low tower, vaulted in two stories, called Cobb Hall, from being the place for floggings. It possibly was built by John of Gaunt, who was custodian of this castle. A second Norman tower-gateway, with its barbican partly existing, breaks the long line of the western wall at its northern end. There are two great artificial mounds on the south side of the castle (as a rule there was only one), one possibly being British. The larger one is about 50 feet high and 100 feet in diameter at its summit. The keep of Norman masonry which stands on it is only a shell, many-sided, the wall about 8 feet thick and 20 feet high. The other mound, of the same height but half the diameter, supports the Observatory Tower, so called from the modern round turret surmounting it which was built as an observatory. This tower must have been of importance, as it commands the main street coming up the Steep Hill into Bailgate. In spite of the long stretch of wall and the few towers thereon, the castle proved in the course of its history to be a hard nut to crack, having only twice been fairly captured in the course of many sieges and attacks.

Eight miles south of Lincoln is Somerton Castle, which was built in 1281 by Anthony Bek, Bishop of Durham. It comprised within its walls a quadrangular area 330 feet long from north to south by 181 feet from east to west. At each angle was a large circular tower. The south-eastern tower, 45 feet high, still exists in a fairly good condition, and an Elizabethan manor house joins on to it. The castle was further defended by two moats. The most interesting piece of history in connection with the castle is the fact that from July 1359 to March 1360 it was the residence of King John (le Bon) of France and his son, Prince Philip, that monarch having been defeated and captured by the Black Prince at Poitiers in 1356.

About 20 miles south-east of Lincoln, close to the river Witham, is Tattershall Castle. It exemplifies well the change that was gradually coming over the country with the introduction of gunpowder, that the nobleman's castle was more and more becoming the nobleman's palace, for in this case the windows of the exposed side of the keep are just as large, as decorated, and as beautiful as on the less exposed sides. The original castle was built after the year 1230. Nothing except the large outer moat and the earthworks inside it remain. The existing portion of the castle represents the keep of earlier days and is a large quadrangular tower 112 feet in height built of small red bricks, with patterns externally in blue-black brick, probably all of local manufacture. The windows, battlements, and fireplaces are of stone, almost certainly from the Ancaster quarries. The fireplaces, which after having been torn out and sold are now replaced, are very beautiful and heraldically interesting. The builder of this splendid specimen of brickwork was Ralph, third Baron Cromwell, who was King Henry VI's Lord Treasurer from 1433 to 1443, Master of the Royal Mews, and Royal Falconer. There is, however, but little history connected with the castle. It sustained

Tattershall Castle

some damage in the great Civil War, after the battle of Winceby most likely, when the Royalists were badly beaten. It was the only fortified place in the county which was garrisoned by the Parliament in 1648, when Lincolnshire was attacked by Royalists from Pontefract, all other defensible places having had their walls, towers, and ramparts "slighted." Its last inhabitant was in the early years of the last century—a pensioner who lived in the gallery in the eastern wall to be ready to light a beacon in the south-east tower in case of invasion.

Kyme Tower, between Sleaford and the Witham, is all that remains of an important castle built by the Umfravilles, Earls of Angus. It consists of three stories with a turret stair, and is groined at the top with fan tracery, springing from a central pillar. It probably was only used as a place of safe retreat, not for living in, as there are no traces of fireplace or floors or chimneys throughout the building.

Of other castles, strong and famous in their day, only the mounds and ditches remain. Such is the case with Sleaford, built by Bishop Alexander of Lincoln about the year 1130. Here King John spent the night of the 14th October, 1216, on his last journey from Kings Lynn across the Wash, when his baggage was lost. Next day he rode through Hough to Newark, and died there on the 18th of that month. Bishop Fleming died here in 1431. Probably the great Civil War would be responsible for its having been "slighted," as it was termed, i.e. made almost impossible of defence.

Folkingham, once the property of Gilbert de Gaunt,

who rebuilt Bardney Abbey, and later of the powerful
Beaumont family, shows only a mound (on which a
gaol, now closed, stands) and a fairly well marked inner
moat. In Leland's time the castle was ruinous, so the
Cromwellian cannon and the orders of Parliament may
not have had much to do with its downfall.

Just where the Lincolnshire wolds sink into the great
plain of marsh and fen, close to Spilsby, a few mounds
alone represent the site of Bolingbroke Castle, the
birthplace of the son of John of Gaunt and the Duchess
Blanche of Lancaster, afterwards King Henry IV, on
April 3, 1366. It was built by William de Roumare,
first Norman Earl of Lincoln, about the middle of the
twelfth century. It is described by Gervase Holles,
before the great Civil War, as built of soft wold
sandstone in a square, with four strong forts (at the
corners probably), containing many rooms, and occupying
about an acre and a half of ground. Queen Elizabeth
also added some rooms. The castle stood a siege of a
few days by the Earl of Manchester and the Parliamentary
army in October, 1643, but after their complete victory
at Winceby fight it was deserted. The perishable nature
of the sandstone, and possibly some "slighting" after its
capture, have left now not one stone upon another, the
gate having fallen down in May, 1815.

At Castle Carlton are three great artificial mounds
covered with trees, near the church, which with their
moats occupy a space of nearly five acres. On the south
and east of the village is a rampart, 12 feet wide and
5 feet high, and about a mile in length. These moated

mounds are all that is left of the once very strong and important castle which was built by Sir Hugh Bardolph in 1295–1302.

The same fate has overtaken the once strong castle of Bourne (connected with the Wake family), and Castle Bytham, the fortress of the Earls of Albemarle.

21. Architecture—(*c*) Domestic.

Of buildings of Norman date other than churches and castles but few have lasted to these latter days; but there are four of these in Lincolnshire which demand attention. The manor house at Boothby Pagnell is of late Norman date, has a vaulted undercroft, an external staircase, and a very early fireplace. The two Jews' Houses at Lincoln (see p. 103) of about the same date are probably the oldest inhabited houses in England; both are two-storied, have chimney-shafts corbelled out over a round-arched door, and have round-headed windows. St Mary's Guild House, also in Lincoln, has a fine semicircular-headed entrance arch, and a rich cornice of foliage runs along the street front. Within the court is a good Transition Norman house, with two-light windows and a plain Norman fireplace. The houses of the Priest Vicars of Lincoln Minster were built between 1280 and 1398 in collegiate fashion round a court. Several of the houses have disappeared, as has the common dining-hall; in the south house are some beautiful decorated windows. The Chancery, Lincoln, was built about 1316, and the picturesque red-brick front and stone oriel window

were added in the time of Bishop Russell (1480–94).
The finest country house in Lincolnshire undoubtedly is
Grimsthorpe Castle, belonging to the Earl of Ancaster.
At the south-east corner remains one of the original
towers of late twelfth century date; the east, south, and
west fronts were built by Charles Brandon, Duke of

St Mary's Guild House, Lincoln

Suffolk, to receive his brother-in-law King Henry VIII
in 1541; while Sir John Vanbrugh erected the stately
but rather heavy north front in 1722. The Angel
Hotel in Grantham is one of the very few medieval
hostelries in existence, the entrance gateway dating from
the fourteenth century—the corbel ends of the weather

moulding being the heads of King Edward III and Queen Philippa. The rest of the front of the house is about a century later. King Richard III signed the death-warrant for the execution of Buckingham here in 1483.

The Rectory of Market Deeping still contains part of the refectory of a Priory which belonged to Crowland Abbey : it has a fine timbered roof and a beautiful

Grimsthorpe Castle

window of fourteenth century date. Wainfleet School, a good example of brickwork, was built by the Bishop of Winchester (William of Waynflete) in 1484, and greatly resembles the much earlier brickwork of Tattershall Castle already described. The Old Hall at Gainsborough was probably built in the reign of King Edward IV, as the Banqueting Hall with the other timber work is of that

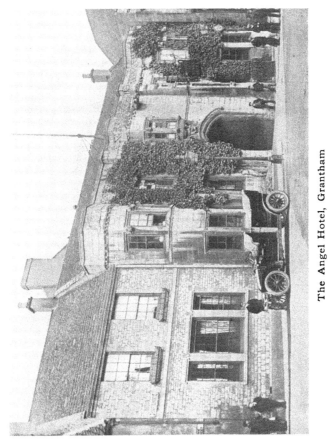

The Angel Hotel, Grantham

Old Hall, Gainsborough

date, as well as its butteries and kitchen, though the oriel window, tower, and gallery were added in the reign of King Henry VII. Some remains of mural paintings perhaps owe their existence to King Henry VIII's visit in 1541, when he stayed here with Thomas, Lord Burgh. Ayscoughfee Hall, Spalding (the residence of Maurice Johnson, now the Museum) is supposed to have been built about 1420, but it has been greatly altered.

The Stone Bow, Lincoln, is a good example of a fifteenth century city gate. The Guildhall is the couple of rooms above the arches, with a fine open-timbered roof. Browne's Hospital, Stamford, the finest of several Hospitals or *Callises*, as they are called, from having been built by Calais merchants of the Staple, was founded about 1480. The half-timbered manor house of Knaith, where Thomas Sutton was born, was built by the Willoughby family of Parham in the early sixteenth century, and the stone manor house at Great Ponton and the fine house at Irnham were erected about the same date.

Scrivelsby Court, the home of the Dymokes, the King's Champions, now much modernised in Tudor Gothic, was probably a good medieval house before the fire in 1761 and has an ancient entrance gateway. Thorpe Hall, built in 1584 by Sir John Bolle, is an interesting Elizabethan house. The ruined manor house of the Jermyns at Torksey, those at Bassingthorpe and North Carlton, the Red Hall of the Digbys at Bourne (now the station-master's house), and the splendid house at Doddington (p. 146), built in 1600, with its spacious windows, flat cornice, and turreted gazebos (much resembling

Knaith Manor House
(the birthplace of Thomas Sutton)

Hatfield House), all date from the reign of Queen Elizabeth. Of later houses Harrington Hall, originally Tudor, was rebuilt in 1678; Belton House, built by Wren in 1659, was added to by Wyatt in 1775; Uffington House was erected in 1688; Summer Castle by Sir Cecil Wray in 1760; Norton Place by Carr, an architect of York, who also enlarged Panton Hall, which was built in 1724 from designs of Hawksmoor, the architect of Queen's College, Oxford. Haverholme Priory is a stately house chiefly of Tudor Gothic, erected in the eighteenth and nineteenth centuries on the site of the Gilbertine monastery, of which some slight remains still exist.

Brocklesby Park is the most important house in the county next to Grimsthorpe Castle. It dates from the eighteenth century, with large additions at the beginning and end of the last century. Langton Hall by Spilsby, where lived Bennet Langton, Dr Johnson's friend, built in 1866 in Elizabethan style, is the fifth in the same park.

In the villages on or near the Cliff, where stone was easy to be obtained, there occur smaller stone houses, with high gables and mullioned windows, either dating from the sixteenth and early seventeenth centuries or, as in the Cotswolds, perpetuating almost to our own time the ancient, excellent fashion of building. Such houses can be found in Navenby, Leadenham, Caythorpe, and Brant Broughton. Many of the villages are mainly composed of stone cottages, originally perhaps of timber and plaster, of which a few remain here and there with their picturesque thatched roofs, though in some cases

Doddington Hall

these roofs are being replaced by that modern abomination, galvanised iron. But the greater number of cottages in the county are of brick, chiefly red; the yellowish white, so common in Cambridge, being fortunately not much made in Lincolnshire.

22. Communications: Past and Present.

The Roman roads have been already considered in the chapter on Antiquities, so it may suffice to say that they generally run in this county at a distance from, and apparently independent of, the villages along their track. This may be due either to these villages having been built in the wooded districts purposely away from the roads for security (if they date from later than Roman times) or to the directness of the Roman roads, which merely joined the various stations together for military purposes. In Robert Morden's map of the county, published in Thomas Cox's *History of Lincolnshire* in the years 1720 to 1731, three roads are marked which enter the county from the south:—(1) the present Great North Road from London and Stamford passing by South Witham to Grantham (this is part of the Ermine Street) and thence to Newark; (2) another road from London entering the county at Market Deeping, thence by Bourne and Folkingham to Sleaford (also a Roman road thus far) and from Sleaford by Leasingham past Dunston Pillar to Lincoln; lastly (3) a road from Stilton entering the county at Crowland, and running through Cowbit, Spalding, Gosberton, and Kirton

Ermine Street

(*The Roman Road leading north from Lincoln*)

to Boston. The Fosseway from Newark to Lincoln is also well marked. From Lincoln the Ermine Street goes as far as Hibaldstow and then becomes the existing road to Brigg and Barton-on-Humber, whence the map quaintly directs "to Flamborough"! Another Roman road leads eastwards through Welton, Snarford, and Market Rasen to Grimsby.

In the earlier maps of the county such as that by J. Hondius in 1610, and that by Blaeu in 1645–50, no roads, unfortunately, are marked. The hill at Lincoln in coaching days must have been always difficult and dangerous. In a map dated 1819 and corrected to 1848, the steep ascent from Hungate through Michaelgate to the South Roman gate is marked "old coach road," and was probably preferable to that going straight up the Steep Hill. A hundred years ago there was one mail coach from London which arrived in Lincoln every afternoon between four and five o'clock and set off for Barton about three-quarters of an hour after its arrival. There was also a light coach which passed through Lincoln morning and evening from and to London and Barton. A coach also started every morning at nine o'clock for Newark and Nottingham, and another left Nottingham every morning and arrived at Lincoln about half-past five the same day. Waggons for the conveyance of goods from Lincoln to London started every Monday and Friday and arrived there in four days, a distance of 134 miles, and returned in the course of a few days, " so that," in the proud words of a contemporary guide book, " there is a regular communication between this place and the capital." Two waggons seem to have carried goods to Brigg and Barton, but how frequently does not appear, and the Sheffield carriers arrived at the Old Crown every Thursday evening and set off again on Friday forenoon. A waggon from Louth also arrived at the Crown on Thursday evening and left again next forenoon.

This stands out in wonderful contrast to the present conditions, when there are 34 trains to London from Lincoln, and 26 trains from London to Lincoln in the day.

Owing probably to its geographical situation, Lincolnshire is not much affected by the main lines of the principal railway systems of this country. That of the Great Northern Railway enters it at Tallington, crosses a little tongue of Rutland at Essendine (whence is a branch line to Stamford) and running north-west to Grantham attains its highest elevation (370 feet) on the east coast route between London and Berwick as it passes through Stoke Rochford. The 105½ miles from London to Grantham, often a non-stop run, is timed to take under two hours. From Grantham the line runs to Barkston, and leaves the county between Claypole and Balderton, just before reaching Newark. Communication with Lincoln is secured by a branch line from Barkston, linking up the villages just on or west of the Cliff; and from Honington another branch line passes through the gap in the Cliff at Ancaster to Sleaford and Boston. At an early date in the construction of railways it seemed possible that Stamford might be the site of the Great Northern Railway's chief repairing and building works, combining the junctions

Crowland Abbey Bridge

of Peterborough and Doncaster. Fortunately for the antiquarian and picturesque interest of the town, this did not take place. An important cross-country line owned jointly by the Great Northern and Midland runs from Melton Mowbray to Bourne, Spalding, Holbeach, and Kings Lynn. The Great Eastern may be said to have one main line through Lincoln, as it comes north from London through Cambridge, Ely, March, Spalding, and Sleaford, and runs from Lincoln to Gainsborough, Doncaster, and York. Boat trains from the north and the west also pass on this route to Harwich for the continent.

The Great Central Railway (in its earlier days the Manchester, Sheffield, and Lincolnshire) connects Doncaster with Crowle, Scunthorpe, Barton-on-Humber, New Holland (for the ferry to Hull), and Grimsby, and has lines from Lincoln and Gainsborough to the same places. It has also taken over the line of the incorrectly named Lancashire, Derbyshire, and East Coast Railway, which runs from Chesterfield through important collieries and the "Dukeries" to Lincoln. It does not exist in Lancashire and never got to the east coast, where at Sutton it was proposed to construct a dock.

From Grimsby and Cleethorpes the Great Northern Railway runs to Peterborough through Louth, Alford, Boston, and Spalding. There is also a branch line to the coast by Theddlethorpe, Mablethorpe, Sutton, and Willoughby; a short line to Spilsby, and another to Skegness; while Boston is connected with Lincoln by a line following the course of the Witham.

The record of the canal system of Lincolnshire is

rather a melancholy one. From the junction of the Witham with the once great Mere at Lincoln, now shrunk to small proportions and called Brayford (Broad ford ?), the Romans constructed a canal—the Fossdyke —westwards to the Trent at Torksey. King Henry I is recorded by Hovenden to have made a long canal from Torksey to Lincoln by digging (though it is almost certain that he merely cleared out a pre-existing dyke, as the presence along its course of so many villages with Scandinavian names testifies). By turning into it the water from the river Trent he made a passage for shipping. This canal was presented by King James I to the City of Lincoln, who granted it on a long lease in 1740 to Mr Richard Ellison. It is now under the control of the Great Northern Railway, and a fair amount of traffic is carried on by barges between Gainsborough and Lincoln, and on the Witham to Boston.

The Cardyke (car = fen) has been already mentioned when dealing with the drainage of the county. It reached for 57 miles from the river Nene to the river Witham, though not navigable, and is still useful as a land-drain in several portions of its course. The Louth Canal, the New Cut for the Ancholme river, and the Horncastle Canal have been already noticed. Owing to the railways obtaining the command of most of these canals, so as to be able to extinguish all competition, the traffic on them has become quite insignificant, and some of them are now not navigable. It might have been possible, it seems, to have made them of great use for the carriage of heavy goods, such as coal, etc., and so saved the railways loss of time and overcrowding of their lines. It is possible, however, that with the advent of motor power the canals may again come into use.

23. Administration and Divisions.

The general business of our country is watched over and administered by the Houses of Parliament. In the Upper House a county such as Lincolnshire is represented by the Peers who have estates in the county, and the Lord Bishop of the Diocese, if it has come to his turn to take his seat in that House, while to the Lower House she sends 11 representatives, being one each for the City of Lincoln, the towns of Boston, Grantham, and Grimsby, and the county divisions of Brigg, Gainsborough, Horncastle, Louth, Sleaford, Spalding, and Stamford. The domestic affairs of the county are administered by three great local bodies, called County Councils, of which there is one for each Part of Lincolnshire, which also possesses its own treasurer, magistrates, Quarter Sessions and Clerk of the Peace (who is also Clerk of the County Council). The Lindsey County Council consists of a chairman, vice-chairman, 16 aldermen and 48 members, and meets at Lincoln. The Kesteven County Council consists of a chairman, vice-chairman, 16 aldermen and 48 members, and meets at Grantham and Sleaford alternately. The Holland County Council consists of a chairman, vice-chairman, 16 aldermen and 42 members, and meets alternately at Boston and Spalding. The

Councils manage the administrative business of the county by a Standing Joint Committee, which among other duties has the appointment of Clerk of the Peace and the control of the police. There are about 300 county policemen, under the charge of a Chief Constable. Before this system was established each of the smaller divisions of the county had its own High Constable, who was responsible for the policing of his own area[1].

There is a large prison at Lincoln, mainly for the county of Lincolnshire, which is managed entirely by the Prison Commission of the Home Office, under the Home Secretary, but a Visiting Committee from the county and from Lincoln attends once a month to investigate complaints and to deal with any specially refractory prisoner.

There are also Urban and Rural District Councils, and Parish Councils. All this is not unlike the form of government in Saxon times, when there was a king and a sort of Parliament, the Witenagemot, and lesser councils for the counties and various divisions down to the villages. The name applied to the larger divisions is in Lincolnshire (as in Yorkshire and Nottinghamshire) wapentake (i.e. "a weapon touching," in acknowledgment of fealty, and hence the district affected by the ceremony) and many of them bear evidence of their Scandinavian parentage. Lindsey has 13 wapentakes, 2 hundreds (the Saxon and more common

---

[1] Probably a vestige of this is found in the existence of this office at Lincoln, where the Sheriff, after his term of office has expired becomes the High Constable for the following year, but his only duty is to determine the date of the Statutes or Hiring Fair.

equivalent, meaning the area containing 100 families), and 2 sokes (literally "a seeking into," hence the precinct within which the right of hearing suits existed). Kesteven has 9 wapentakes and 1 soke, while Holland has 3 wapentakes.

At the official head of the county comes the Lord-Lieutenant, who is usually a nobleman or great landowner appointed by the Crown, and who was, in past times, responsible for the troops of Militia, etc., within his sphere of action. Under him are several Deputy-Lieutenants who will probably take much more active interest in the Territorial forces than has been the case for many years past. The representative of the King and his officer in the county is the High Sheriff, whose name is pricked by the King every November out of a list furnished by the Chancellor of the Exchequer. The Assizes are held three times a year at Lincoln for the City and County, and Lincoln, Grantham, and Grimsby have each a special judicial officer, the Recorder, a barrister of eminence, who holds Sessions to try prisoners sent to him by the magistrates four times a year. There are also Quarter Sessions for the three Parts of the county, and Petty Sessions are held frequently in various places for the County and City Justices of the Peace to try all offenders against the law.

Lincoln City must have long experienced a kind of domestic Home Rule, for when she was a Roman colony there was much of local rule, and still more when, as a member of the Danelagh (with Stamford), she had her own Earl with his separate "host," while twelve lawmen administered Danish law and a common Justice Court existed for the whole confederacy. From King Richard I came the right to elect the City's Provost (there were two allowed in 1227) and a Mayor was evidently elected—the Provosts becoming Bailiffs early in the thirteenth century. The first date relating to a Mayor of the City is 1210. King Henry IV gave the city the privilege of electing two Sheriffs in the place of the Bailiffs, as he had granted Lincoln the privilege of being styled "The County of the City of Lincoln." It is now governed by a mayor, 6 aldermen and 18 councillors, and has its own sheriff and coroner, the latter a very ancient officer whose duty it is to investigate into the cause of all doubtful or suspicious deaths.

Boston was incorporated by King Henry VIII, Grantham by King Edward IV, Grimsby by King John, Louth by King Edward VI, and Stamford by King Edward IV. Louth and Grimsby also possess the additional office of High Stewards.

There are 18 Poor Law Unions, under Boards of Guardians, to manage the workhouses and relieve the poor by specially appointed officers. The Registration County of Lincolnshire does not coincide with the Geographical County, as it includes several parishes from neighbouring counties, and has several of its own parishes allotted to other counties.

The ecclesiastical government of Lincolnshire is in the hands of the Lord Bishop of the Diocese, who is assisted by two Archdeacons, of Lincoln and of Stow, who each supervise 21 associations of parishes over which a Rural Dean presides. In many cases the names of these Rural Deaneries and their areas are almost identical with the secular divisions (wapentakes and hundreds) already mentioned, but the ecclesiastical names have frequently kept nearer to the original forms. There are 582 parishes in the diocese, but in a number of instances two or more are joined together, either from the loss of one church, as at Mablethorpe, Claxby-Pluckacre, and Fordington, or on account of the smallness of the stipends attached.

24. The Roll of Honour.

"This Shire triumpheth," remarks John Speed, "in the birth of King Henry the fourth, at Bullingbrooke borne." He was son of John of Gaunt, "Time-honoured Lancaster," who lived in Lincoln, where a small part of his palace still exists, and whose third wife and daughter, Katherine Swynford and Joan Countess of Westmorland, are buried in the Minster. He goes on to say with less accuracy, "but may as justly lament for the death of King John, herein poysoned by a monk of Swynsted Abbey" (as noted above King John died at Newark Castle); "and of Queene Eleanor, wife to King Edward the first, the mirrour of wedlocke and loue to the Commons, who at Harby[1] ended her life."

Of early Saints Lincolnshire can claim two as belonging to her perhaps by birth, certainly by life—St Botolph, who died about 680, and whose memory survives in Boston—Botolph's town, and St Guthlac, who died in

---

[1] Just outside the Lincolnshire border, in Nottinghamshire.

John Wesley

713, and in whose honour Crowland Abbey was founded.
In later times St Gilbert of Sempringham (1083–1189 ?),
Lincolnshire born and bred, founded the only monastic
order—that of the Gilbertines—which originated in this
country. Mention has already been made of the many
founders of Colleges, so that out of the great roll of
Bishops of Lincoln room can be found here only for St
Hugh (1186–1200); Sanderson, whose life was delightfully written by his friend, Izaak Walton; and in these
later years Christopher Wordsworth, scholar and divine,
and the greatly beloved and saintly Edward King.
Archbishop Whitgift (1530–1604) was born in this
county, at Grimsby, and Daniel Waterland, theologian
(1683–1740), at Walesby, of which place Robert Burton,
author of *The Anatomy of Melancholy* was once rector.
William Paley, author of the famous *View of the Evidences
of Christianity*, probably wrote his *Natural Theology* when
Sub-Dean of Lincoln.

John Cotton, nonconformist divine, was Vicar of
Boston for several years before leaving for Trimountain,
Massachusetts, afterwards called Boston. And the Lincolnshire Boston was the native place of John Foxe
(1516–1587) the martyrologist. In 1703 there was born
in Epworth Rectory one who made a vast change in the
religious world, especially of that portion, unlike himself,
outside the Church of England. This was John Wesley.
His brother Charles, the divine and hymn-writer, was
born there in 1707. One of Lincolnshire's proudest
boasts should be that the great genius, Sir Isaac Newton,
was born at Woolsthorpe (in 1642) and educated at the

neighbouring grammar school of Grantham, where there is a statue of him by Theed. Another mathematician,

Sir Isaac Newton
(*From the statue by Roubiliac in the chapel of Trinity College, Cambridge*)

George Boole, author of *The Laws of Thought* (1815–1864), was born in Lincoln.

On August 6, 1809, the poet, Alfred Tennyson, was born in Somersby Rectory. He was educated at Louth grammar school. His earlier work reveals much of the influence on his receptive mind of the varying moods of river and marsh and fen. His statue by G. F. Watts is on the north-east side of the Minster Green at Lincoln. His elder brothers, Frederick (author of *Days and Hours*) and Charles, who took the additional name of Turner, were also poets. Boston was unusually favoured in the early years of the nineteenth century, as James Westland Marston, poet, dramatist, and critic, and father of the blind poet, Philip Bourke Marston, was born there in 1819; Jean Ingelow, the poetess, in 1820; John Conington, the famous classical scholar in 1825; and Herbert Ingram, the founder of the *Illustrated London News* in 1811. Thomas Cooper, the chartist (author of *The Purgatory of Suicides*), lived for the later part of his life at Lincoln, where he died in 1892. Mrs Centlivre, actress and dramatist (1667–1723), was born at Holbeach, while Thomas Heywood, dramatist and poet (died 1650?), avers himself a Lincolnshire man in his commendatory verses to James Yorke's *Union of Honour*, a book of heraldic use, by a Lincoln blacksmith. Robert Mannyng, or Robert de Brunne (1288–1338?), poet, was born at Bourne; Henry, Archdeacon of Huntingdon, the historian (1084–1155), was buried at Lincoln; and to turn to lighter literature, John Sheffield, 1st Duke of Buckinghamshire, the friend of Dryden and Pope, had his seat at Normanby, the home of Sir Berkeley Sheffield. Bulwer Lytton, who was M.P. for Lincoln, is reported to have written *A Strange*

Alfred, Lord Tennyson

Story in the grounds of the old Palace, and the scene of the story is laid in Lincoln.

Among antiquaries respect will always be paid to the names of Maurice Johnson (1688–1755), founder of the Gentlemen's Society (the oldest antiquarian society in England) at Spalding, his birth-place; of William Stukeley (1687–1765), author of *Itinerarium Curiosum*; of Sir Charles Anderson of Lea (1804–1891), author of the charming *Lincoln Pocket Guide*; of Edward Trollope (1817–1893), Bishop of Nottingham; and of Edmund Venables (1819–1895), Precentor of Lincoln. To both these latter Lincoln and Lincolnshire archaeology owes very much.

Of explorers this county has produced many of marked distinction, beginning with a typical "scout," Captain John Smith of Willoughby (1580–1631) one of the Virginian colonists, who was rescued from the Indians by their Princess Pocahontas; the traveller, Fynes Moryson (1566–1617), of Cadeby (?); Sir Joseph Banks (1743–1820), one of the most distinguished of naturalists and President of the Royal Society, who lived at Revesby and was companion of Cook in his voyage in the *Endeavour* round the world; Matthew Flinders of Donington (1774–1814), explorer of Australasia; and, best known of all, Sir John Franklin (1786–1847), "heroic sailor soul," the Arctic explorer, whose statue is in Spilsby market place.

Among famous soldiers[1] may be mentioned Sir John

---

[1] The history of Hereward the Wake and his connection with this county is, unfortunately, almost entirely legendary.

Statue of Sir John Franklin

(*In Spilsby market-place*)

Bolle of Haugh (where he is buried) and Thorpe Hall, the hero of the *Spanish Lady's* ballad; Bartholomew, Lord Burghersh (died 1369), who fought at Crecy, and whose praise as a brave and gallant knight is enshrined in Froissart; Peregrine Bertie, 11th Lord Willoughby de Eresby (1555–1601), the hero of the ballad, *The bravest man in battel was brave Lord Willoughby*; and Francis Willoughby, 5th Lord Willoughby of Parham (1613–1666), a great Parliamentary general who afterwards returned to his allegiance to his King.

At the manor house at Knaith near Gainsborough was born in 1532 Thomas Sutton, a member of a prominent Lincolnshire family, who after service in the army as Surveyor of Ordnance, amassed great wealth and founded the Charterhouse Hospital and School in 1611, in which year he died. A portrait of him hangs in the Guildhall at Lincoln. The greatest name among Lincolnshire statesmen is that of William Cecil, Lord Burghley (1520–1598), born at Bourne, Queen Elizabeth's chief minister. Sir John Cust (1718–1770) and Sir Robert Sheffield (in 1510 and 1512) were Speakers of the House of Commons.

In the world of commerce it is interesting to note that our county supplied no less than nine Lord Mayors of London between the years 1470 and 1632.

For many centuries Scrivelsby has been associated with the family of Dymoke. Sir John Dymoke by his marriage with the heiress of the Marmions succeeded to the manor and was made King's Champion at the coronation of

Thomas Sutton

Richard II., an office ever after retained by the family, though no longer exercised.

Of artists the county certainly has had few : William Hilton (1786–1839), Royal Academician and historical painter, was born in the gatehouse of the Vicar's Court at Lincoln, the son of a portrait painter of that city. His sister married Peter de Wint, who has shown his love for the city and county in many beautiful water-colours. A monument to the two painters is in Lincoln Cathedral. The splendid illustrations of Roman tesselated pavements produced by William Fowler (1761–1832) must not be left without mention.

In the medical world Francis Willis (1718–1807) attended King George III in his attacks of mental derangement, and John Conolly (1794–1866), born at Market Rasen, and Edward Parker Charlesworth (1783–1853) were also distinguished for their treatment of insanity, the method of non-restraint being introduced by the latter, whose statue stands just south of The Lawn—a private asylum at Lincoln.

One of the most distinguished musicians of the sixteenth century was William Byrd, organist of Lincoln Minster from 1563–1572 ; and John Reading, the composer of *Adeste fideles* (O come, all ye faithful), was Master of the choristers there in 1702.

25. THE CHIEF TOWNS AND VILLAGES OF LINCOLNSHIRE.

(The figures in brackets after each name give the population in 1911, and those at the end of each section refer to the pages in the text.)

Alford (2394), a market town at the junction of the Wold with the Marsh, with a fine church of Decorated date and handsome chancel screen. (pp. 12, 39, 62, 75, 152.)

Algarkirk (485), 6½ miles south-south-west of Boston, has a very fine cruciform church chiefly of Early English and Decorated date. Woad is grown in this parish.

Ancaster (536), 8 miles north-east of Grantham, was a Roman station on the Ermine Street (probably Causennae), the fosse or ditch is clearly traceable. Quarries for Ancaster stone are worked in this parish and in the neighbouring one of Wilsford, where is a picturesque manor house built by Sir Charles Cotterell, a scholarly courtier of Charles II. (pp. 11, 19, 31, 86, 109, 115, 117, 134, 150.)

Bardney (1302), a thriving town 9 miles east of Lincoln, originated with the famous abbey dedicated to St Oswald, of which the ground plan has been worked out. Aethelred resigning his crown became abbot here, and a mound still called King's Hill may be his burial place. There is a junction here connecting the Lincoln and Boston line with Louth. (pp. 12, 137.)

Barrow-on-Humber (2734), includes New Holland, whence there is a steam ferry to Hull. A Saxon monastery was founded here by St Chad, and there is a large series of earthworks, "The Castles." At the neighbouring hamlet of Burnham to the south was possibly the site of the Battle of Brunanburgh in 937.

Barton-on-Humber (6673), 6 miles south-west of Hull, a prosperous market town and small port, which had a market and ferry at the time of Domesday survey, with a Saxon church (St Peter) having a very early tower and handsome Decorated nave and chancel, and a second fine Early English church (St Mary's), almost entirely in that style. The chief trades of the town are malting, and brick, tile, and cement making. (pp. 11, 32, 33, 35, 42, 49, 74, 96, 118, 149, 150, 152.)

Baumber (354), a little village 4 miles north-west of Horncastle, chiefly notable for its training stables; Galopin, winner of the Derby in 1875, was bred here.

Boston Grammar School

Billingborough (964), a large village 3 miles east of Folkingham, with fine church mainly Decorated, a Tudor Hall, and abundant springs.

Billinghay (1288), a large village 8½ miles north-east of Sleaford, includes Dogdyke eastwards on the Witham, where is a piece of genuine untouched Fen, and Walcot, westwards, with the well-known Catley Abbey springs of natural seltzer-water.

Boston (16,673) (St Botolph's town), has a very ancient history; its importance as a port and fishing-centre has already been described. Many of the "Pilgrim Fathers" hailed from this place, and re-christened the town of Trimountain after it. The church (St Botolph's) is a little less in area than St Nicholas, Yarmouth, and is a splendid specimen of a first rank parish church. It is almost entirely of Decorated date, except the grand Perpendicular lantern tower, 272 feet high, known locally as Boston Stump. There is a picturesque half-timbered house, Shodfriar's Hall, a fifteenth century red brick Guildhall, and a sixteenth century Grammar School, also in red brick. The neighbouring church of Skirbeck has a grand Early English nave. (pp. 4, 6, 12, 15, 19, 20, 32, 34, 38, 47, 50, 70, 73, 74, 82, 89, 90, 93, 94–97, 110, 123, 124, 150, 152, 153, 154, 157, 158, 160, 162.)

Bourne (4343) is a nice market town 9½ miles west or Spalding, with a powerful spring, from which it gets its name. The parish church is the nave of the monastic one, with late Norman arcades. Bourne was the seat of the Wake family, though the connection with Hereward is very obscure. The castle and station-master's house have been previously mentioned. (pp. 11, 20, 128, 138, 143, 147, 152.)

Brant Broughton (531), a village 3 miles west of Leadenham, has one of the most perfect parish churches in the county, both outside and inside. It is chiefly of the Decorated period, though the nave, aisle, arcade, and chancel arch are Early English. The interior decoration, stained glass, screen-work, font and cover, and iron-work, is all of a very high order of merit. Here Bishop Warburton was rector and wrote his *Divine Legation of Moses*. (pp. 82, 145.)

Brigg or **Glanford Brigg** (3343), a thriving market town 23 miles north of Lincoln on the Ancholme, over which is the bridge which gives the place its most used name. On the site of the gasworks was discovered a prehistoric boat, 48 feet long, cut out of the trunk of an oak. (pp. 15, 16, 31, 112, 120, 149, 150, 154.)

Burgh (937), a small market town 6 miles east of Spilsby, on an outlying hill in the Marsh, was a Roman station; has a fine Perpendicular church, and a Missionary College founded thirty years ago. (pp. 11, 16, 32, 115.)

Caistor (1544), a small market town 7½ miles east-south-east of Brigg, was a Roman station, and possesses part of its walls and ditch. It is situated on a western spur of the Wolds. The church tower is probably Norman, the nave Early English. A gadwhip is kept here, to which a purse was attached containing 30 silver pennies and four pieces of wych elm, and on Palm Sunday was cracked by a man from Broughton during the reading of the first lesson. He knelt before the officiating minister when he began to read the second lesson, waved it three times over the minister's head, and held it over him till the lesson was finished, and then deposited it in the seat of the Lord of the Manor of Hundon. Lands in Broughton (near Caistor) were held by this ceremony, which has been discontinued for 70 years. (pp. 6, 32, 62, 99, 108, 115.)

Cleethorpes (21,417). This favourite watering-place is situated about 3 miles south-east of Grimsby (with which it is practically continuous), on the south shore of the Humber and facing the North Sea. It is frequented by many thousands of visitors, for day trips or longer, during the summer. About a mile inland is the church of Old Clee, which has a Saxon tower, Norman nave, and Transitional Norman north transept. A tablet records the dedication of the church (perhaps the chancel and transepts) by St Hugh, Bishop of Lincoln, in 1192. (pp. 42, 47, 49, 89, 152.)

Crowle (2853), a small market town in the Isle of Axholme on the Yorkshire border with a station on the light railway from Haxey to Goole. The church, dedicated to St Oswald, has the stem of a Saxon cross with runic inscription in the arch of the tower, which is Norman with Perpendicular top. It has also a fine Norman doorway. (pp. 69, 152.)

Epworth (1836) is a market town in the Isle of Axholme about 6 miles south of Crowle, and has a good church with a fine Perpendicular tower. In the old Rectory (which was burnt by the parishioners in 1709) John Wesley was born in 1703 and Charles Wesley in 1708. Samuel Wesley, their father, was rector of Epworth for 39 years, and is buried in a tomb in the churchyard on the south side, from which his son John used to preach. (pp. 14, 69, 160.)

Frieston (1024) is a large village 3 miles east of Boston. The parish church is the nave of the Priory Church, of Transition Norman; with fine tower, clerestory, and aisle of Perpendicular date. The beautiful font cover is of the same date. (p. 128.)

Friskney (1373) is a large village on the coast, 14 miles north-east of Boston, with a fine (mainly Perpendicular) church, which is especially remarkable for its wall paintings of early fifteenth century date.

Fulbeck (611), 9 miles north of Grantham, is situated on a well-wooded lower rise of ground between the Cliff and the plain and is perhaps the most delightful village in the county. The Hall has been a seat of the Fane family for nearly 300 years. (pp. 29, 54.)

Gainsborough (20,587) is a large market town on the banks of the Trent, 15½ miles north-west of Lincoln. Here, in 868, King Alfred is recorded to have married Ealswitha, daughter of Ethelred, chief of the Gaini (whence the town gets its name).

The earthworks occupied by the Danes, the interesting Old Hall, the history of the place during the great Civil War, the malting, seed crushing, and agricultural implement works, have all been alluded to earlier in this book. George Eliot describes the town under the name of St Ogg's in the *Mill on the Floss*, and the Bazaar took place in the Old Hall. The "eagre" or bore on the river reaches a little above Gainsborough, as do the ordinary tides in the Trent, the Spring tides reach about 10 miles further up the river, to Newton, where the picturesque red cliffs

Woolsthorpe Manor House

(the birthplace of Sir Isaac Newton)

are much visited by picnic parties in the summer. (pp. 6, 7, 11, 15, 18, 24, 26, 62, 78, 80, 82, 97, 99, 101, 107, 110, 111, 114, 117, 124, 140, 152, 153, 154, 166.)

Great Gonerby (1296), two and a half miles north-north-west of Grantham, was well-known as affording the steepest hill on the Great North Road, and as such is mentioned in Scott's *Heart of Midlothian* (chap. 29). (p. 29.)

Grantham (20,070), a large and increasing market town, and a parliamentary and municipal borough, is situated 24 miles

Grantham: the Old Grammar School

south of Lincoln and 105½ from London. It is notable for its splendid church with the finest spire in the county, and the

Angel Hotel, one of the three remaining medieval hostelries in the country, which with the large agricultural implement and other engineering works, have been already mentioned. Sir Isaac Newton was born at the neighbouring manor house of Woolsthorpe and educated at the Grantham Grammar School. (pp. 6, 10, 11, 17, 19, 71, 80, 82, 86, 109, 110, 115, 121, 122, 124, 139, 147, 150, 154, 156, 157, 161.)

Great Grimsby (76,659) has been already dealt with as far as its gigantic fishing interests and the port generally are concerned. The name is traditionally derived from that of a fisherman "Grim," who (according to the poem of *Havelock the Dane*) rescued a Danish chief's son Havelock from drowning and was rewarded by a gift of this Danish port. It is a Borough by prescription. King Richard I held a parliament here, and King John visited the town twice, and gave its first charter. One only is left of two fine churches (St Mary's fell into ruins in the sixteenth century), St James, the parish church, which belonged to Wellow Abbey. Of early thirteenth century date are nave, south door, porch, transept and part of the chancel—the tower was rebuilt in 1365. The Town Hall, opened in 1863, is of white brick and stone in Italian style; the Corn Exchange in red brick dates from 1854. A fine bronze statue of the Prince Consort stands opposite the Royal Hotel. Archbishop Whitgift and Gervase Holles, author of valuable notes on Lincolnshire churches in the seventeenth century, were natives of Grimsby. (pp. 4, 14, 24, 34, 42, 43, 47–50, 62, 73, 87, 88, 90–93, 97, 120, 149, 152, 154, 156, 157, 160.)

Haxey (2035), a small town 3 miles south of Epworth, 7 miles north-west of Gainsborough, on the G.N. line from Lincoln to Doncaster, and on the light railway to Goole. It was the capital town of the Isle of Axholme. The church is handsome externally, of Perpendicular date with a good tower, "which has a ring of six bells with chimes of exceptional sweetness"

(Jeans). Here on January 6th occurs a kind of football match, called "throwing the Haxey hood," a piece of sacking closely tied up serving for the hood, while 12 "Boggins" and a "Fool" (who proclaims the amount of refreshment offered by the various public-houses of the town to the man who conveys the hood thither) try and keep the hood from leaving a field, the rest of the players striving to capture it. (p. 71.)

Heckington (1666), a large village 5 miles south-east of Sleaford with a magnificent Decorated church which has a beautiful Easter sepulchre and sedilia, in the chancel made by the same workmen as the Easter sepulchres at Navenby, Hawton near Newark, and the Minster chancel screen. (p. 122.)

Holbeach (5052), a small town 8 miles east of Spalding, with an enormous parish, containing 21,000 acres of land, and 14,000 of water. Henry Rands, Bishop of Lincoln (1547–1551), who alienated most of the episcopal manors to the king, and Stukeley the antiquary, were natives of this place. The church is large and dignified, of the latest Decorated date, with a spire 180 feet high; the north porch has curious circular battlemented turrets, like those at the east end of the Lady Chapel at Grantham Church. (pp. 69, 152, 162.)

Horncastle (3900) is a market town at the foot of the Wolds, 21 miles east of Lincoln, on an angle of land between the rivers Waring and Bain (the latter giving rise to the Roman name, Banovallum), hence the more modern name. Portions of the Roman walls still exist. There is a good trade carried on in corn, coal, malting and brewing, and the largest horse fair in the kingdom. The manor was possessed for some centuries by the Bishops of Carlisle, as a safe retreat from the incursions of the Scots. The church is fine, with some Early English portions, but the nave and aisles are of Decorated date, and the chancel Perpendicular. A monument to Sir Lionel Dymoke, King's

178 LINCOLNSHIRE

Champion in 1519, another to Sir Ingram Hopton, who nearly captured Cromwell at Winceby fight, in 1643, and a number of scythe heads, most probably used in the Lincolnshire rising in 1536, deserve notice. (pp. 6, 16, 17, 38, 62, 69, 83, 108, 115, 116, 153, 154.)

Kirton-in-Holland (2444) a large village 4 miles south-south-west of Boston, has a fine church which was barbarously treated in 1804, when its central tower and a fine transept were pulled down, and the chancel shortened. The present west tower was then built. The nave is Early English, with clerestory of good Perpendicular work, and a fine roof.

Kirton-in-Lindsey (1602), a market town and large parish, 18 miles north of Lincoln, has a large church with Early English tower, and north arcade of the nave. This place was given by King William the Conqueror to Bishop Remigius towards endowing his new Cathedral at Lincoln, and remained attached to the sub-dean's office for many years as a "Peculiar." (p. 147.)

Langton-by-Spilsby (168) is chiefly remarkable for the fact that the manor and advowson and the estate have been in the possession of the same family (Langton) since the thirteenth century. The Hall has been rebuilt about four times. Bennet Langton was visited here by Dr Johnson. (p. 145.)

Lincoln (57,285), a city and county by itself, returning one member to Parliament, occupies the brow of a cliff 210 feet high, and extends downwards across the river Witham for two miles. It is the episcopal see of the diocese of Lincoln, and is governed (as it has been since 1210) by a mayor and corporation. Its early history, its extensive Roman remains, and its later history have been already dealt with, as have been also its trade, the Jews' Houses, St Mary's Guild, and the Castle. The Cathedral or Minster, as it is usually called, was first built by the first

Norman Bishop Remigius, and completed in 1092. The west
front of this church and one bay of its nave alone remain.
The three splendid Norman doorways and the lower part of the
western towers and the gables are later additions. The saintly
Bishop St Hugh of Lincoln (1186-1200) began rebuilding in
Gothic style, and much of the present ritual choir and western
transepts was completed before his death. The rest of the

Lincoln Minster, from the North-east

transepts with part of the central towers, the nave (a superb work
of great lightness and elegance) and the Chapter House were
built in the time of Bishops William of Blois and Hugh of
Wells. The great west screen, with the central (Broad or Rood)
tower is due to Bishop Grosseteste, the greatest of Lincoln's
Bishops. The extreme popularity and sanctity of St Hugh's
shrine led to the pulling down of the apsidal end of his choir,

and to the replacing it by a most sumptuous and splendid work of Early Decorated date which was consecrated in 1280, when St Hugh's remains were translated to a new shrine. This was called, from the artistic sculpture in the spandrels, the Angel Choir. The south porch is almost unique in the country, with lavish figure sculptures. The cloisters were built by Bishop Oliver Sutton (1280–1300), and are elegant specimens of Decorated date.

Horse Fair, Lincoln

The great central tower was completed in 1311, when with its spire it was about the highest in Europe, reaching a height of 525 feet. The beautiful chancel screen, with its exquisite side doors (like the king's door at Trondhjem Cathedral), was made about the same time. About 1380 the tabernacled stalls of the choir, according to Pugin "by far the finest in England," were put in. The upper part of the western towers (Perpendicular in style, but blending wonderfully well with the Norman under-structure) dates from 1390. After the Restoration, Sir Christopher Wren completed the north side of the cloisters with an arcade and built a library over it.

A fine statue of Tennyson by G. F. Watts stands on the north-east Minster green. St Mary-le-Wigford, St Peter-at-Gowts, and St Benedict have all pre-Norman or early Norman towers; St Swithin's is the finest new church in the city. There are plenty of recreation grounds around the city, the Arboretum above Monks Road, a well-kept People's Park, Monks Abbey, and the South Park on the slope of Canwick and Crosscliff Hills. Horse-racing is mentioned in the Corporation records in 1597. King James attended races here in 1617. The Lincolnshire Handicap, the first great flat race of the year, was first run in 1849. (pp. 1–3, 6, 7, 10, 11, 12, 14, 15, 16, 18–21, 24, 29, 31, 35–38, 41, 46, 54, 56, 61, 62, 66, 67, 69, 73, 74, 75, 78, 79, 82, 85, 86, 96, 97, 99–111, 112–126, 130–139, 143, 147–158, 160, 162, 168.)

Louth (9880), a municipal borough and market town, deriving its name from the river Ludd, is situated in a valley on the east side of the Wolds, about 15 miles south of Grimsby. It is a handsome, well built, and prosperous town, and a great agricultural centre. The main feature of the town is the large and splendid church of Perpendicular date, with its grand tower and spire, rising to a height of 300 feet. Close to Louth on the Lincoln road is Thorpe Hall, built in 1584 by Sir John Bolle, the hero of the *Spanish Lady's* ballad, who is buried at Haugh. The grammar school had the distinction of educating all the Tennyson brothers, Hobart Pasha, and Governor Eyre of Jamaica. Louth Park Abbey ruins are about 1½ miles away; it was a large Cistercian foundation, originating in a colony of monks who left Haverholme Priory. (pp. 10, 18, 33, 124, 150, 152, 153, 154, 157, 162.)

Mablethorpe (1232) is a pleasantly situated watering-place on the coast 8 miles north-east from Alford, with many resident visitors, and crowds of trippers in the season. The sands are excellent and very extensive at low tides. Mirages are not uncommon. Tennyson, who frequently visited the place in his early years, used to say that nowhere, except on the west coast of Ireland, had he seen such length of wave, and many allusions to this coast and the neighbourhood will be found in his poems. (pp. 38, 43, 44, 49, 50, 99, 152, 158.)

Marsh-Chapel (551) is a large parish 8½ miles south-east of Grimsby, with a spacious Perpendicular church, perhaps the finest in the Marshes, with a good tower and a fine rood screen. (pp. 14, 49, 124.)

Moulton (2226), a large parish 4 miles east of Spalding, has one of the grand series of churches on the road between Spalding and Sutton Bridge which belonged to Spalding Priory. It was built about 1180, the nave showing very Early English work of six bays, the clerestories are Transitional Norman, and there is a Perpendicular chancel, and a fine early Perpendicular tower and spire about 160 feet high. Half a mile north-east is the Elloe stone, which marks the place of assembly for the wapentake.

Pinchbeck (2836), a large village of over 13,000 acres, 2 miles north of Spalding, with a large and fine church with Early English arcade of five bays in the nave, Perpendicular clerestory roof and west tower, and late Decorated chancel. The name Pinchbeck for an alloy of copper and zinc came from Christopher Pinchbeck, its discoverer, but whether he had any connection with the village does not appear. (p. 20.)

Market Rasen (2296), is a market town 14 miles north-east from Lincoln. Two other villages of the same name are close by, Middle Rasen and West Rasen, both with interesting churches. The name comes from the little river Rase. (pp. 15, 16, 18, 149, 168.)

Ruskington (1214), a large village about three and a half miles north of Sleaford, has large building works, and a good Early English church.

Sedgebrook (168), a small village 4 miles north-west from Grantham, has a rather remarkable late Perpendicular church, with the rood loft carried across the aisles. The chapel on the south of the chancel was built by Sir John Markham, a Justice of King's Bench, called "the upright Judge" because he was dismissed from office by King Edward IV after displaying conspicuous fairness in dealing with the trial of Sir Thomas Coke, Lord Mayor of London. (pp. 28, 124.)

Skegness (3775) is a rapidly developing watering-place on the coast, 4 miles east of Burgh, and 5 miles north-east of Wainfleet. It has very good accommodation for visitors, who are very numerous in the summer months. The sands and bathing are good, there are a fine pier and cricket ground, and there are first class golf links at Seacroft, on the sand-hills south of the town. The air is wonderfully pure and bracing. The railway communication is through the Great Northern branch from Firsby, either from Boston or Louth, but a connecting line is being made with the Great Northern Railway line from Lincoln. (pp. 14, 31, 43, 46, 47, 49, 50, 54, 62, 97, 152.)

Sleaford (3808) is a thriving and improving market town, of much importance as an agricultural centre, situated about half way between Spalding and Lincoln on the little river Slea. The railway communications are excellent. The church is one of the first class, has a very early spire, a handsome Perpendicular nave (externally Decorated) and a Perpendicular chancel with the finest oak rood screen in the county. The Castle only remains in the shape of mounds. The large maltings have been already mentioned. (pp. 7, 11, 19, 20, 68, 104, 111, 116, 122, 136, 147, 150, 152, 154.)

Somersby (47), 6 miles north-east of Horncastle, was the birthplace of Alfred Tennyson in 1809. The old Rectory is much the same internally and externally as in the days of the Tennysons, with the quaint Gothic dining-room built by Dr Tennyson, who himself carved the mantelpiece. His grave is in the churchyard. The church is a simple little one of the local sandstone, which has been put in good repair, and a bust (by Woolner) of the poet placed in the chancel in memory of his centenary in 1909. There is a beautiful fifteenth century churchyard cross. (pp. 12, 130, 162.)

Spalding (10,308) is a large market town and port, 14 miles south-south-west of Boston, on the river Welland, which is navigable for vessels of 120 tons, carrying on a trade of coal, oil-cake, and timber. It is also a valuable agricultural centre, and for distributing fruit and vegetables. Scarcely any remains exist of the important Priory which was the richest in the county, after Crowland, at the dissolution. The church is mainly Early Decorated, but has had additional aisles added to the nave, and a south-east chapel. The rood screen is fine and very lofty. Ayscoughfee Hall, now town property, was originally of fifteenth century date; it has been much modernised, but has fine Tudor gardens. Maurice Johnson lived here, the founder in 1710 of the Spalding Gentlemen's Society, which still exists and has established itself in a new house. (pp. 12, 20, 22, 69, 97, 143, 147, 152, 154, 164.)

Spilsby (1464) is a pleasant market town 10 miles east-south-east of Horncastle at the southern extremity of the Wolds, on a branch of the Great Northern Railway from Firsby. The church is chiefly interesting for the series of monuments of the Willoughby de Eresby family, who took their title from Willoughby

Somersby Old Rectory

(Tennyson's birthplace)

near Alford, and Eresby, where was a Hall, about three-quarters
of a mile south of Spilsby. A bronze statue of the Arctic explorer,
Sir John Franklin, who was born here, is in the market place.
(pp. 11, 17, 18, 31, 62, 130, 137, 152, 164, 165.)

St Mary's Church and Hill, Stamford

Stamford (9647), a most charming municipal borough,
situated on the extreme southern border of the county on the
Welland, and with one parish, St Martin's, in Northamptonshire
(as is Burghley House, the magnificent seat of the Marquis of
Exeter, built by Lord High Treasurer Burghley). It has an
ancient history, and was one of the Danelagh towns, representing

the Gyrwas or dwellers in the Fens. It has several fine and interesting churches, among them St Mary's with an Early English tower and spire and All Saints with a Perpendicular spire and excellent brasses. Browne's Hospital has been already noticed, and the gate of Brasenose College (the nose has gone to its namesake at Oxford) still remains as a reminder of the University that might have been at Stamford. (pp. 8, 11, 22, 31, 66, 69, 75, 98, 100, 107, 109, 115, 124, 128, 143, 147, 150, 154, 156, 157.)

Stow (309), six miles south-east of Gainsborough, is notable for its large cruciform church, and is supposed to owe its name to St Etheldreda having made a stay (or stow) here, where her staff budded in the night. Parts of the piers of the central tower are probably earlier than 1020, when the church was rebuilt by Bishop Eadnoth; the nave and upper part of the transepts were due to Remigius, and the chancel and fine western doorway to Bishop Alexander. Here was a manor house of the Bishops of Lincoln, and here St Hugh's favourite swan lived. (pp. 10, 118, 120, 157.)

Sutton-on-Sea (835) is another of the pleasant sea-side resorts of this county about 3 miles south of Mablethorpe, and with much the same attractions. (pp. 34, 44–46, 49, 50, 152.)

Swineshead (1899), a large village about 6 miles south-west of Boston, has a grand church, with Decorated and Perpendicular tower, Decorated nave and clerestory and south aisle. A mile away is the site of Swineshead Abbey, occupied by a farmhouse, where King John lay the night after the disastrous crossing of the Wash, and where he was possibly poisoned by the monks. (pp. 64, 104, 117.)

Tattershall (415), on the river Bain 11 miles north-west of Boston, has the splendid castle already described and a spacious Perpendicular cruciform church with western tower, built of stone by Ralph, Lord Cromwell, left unfinished at his death in 1455, and finished by William of Waynflete, Bishop of Winchester. The stone pulpitum or screen dividing the chancel from the transepts has had two altars on its western face, and has two stone book-rests on the projection over the door into the chancel, for the books of the Gospel and Epistle, which used to be read therefrom. There are very fine brasses commemorating the founder and others in the north transept. The market was gained for the town at the price of a well-trained goshawk in the reign of King John, and the market cross bears the shields of Tattershall, Cromwell, and Deincourt, and was erected about the same time as the castle. No market is now held, but there is a cattle and sheep fair in September. (pp. 17, 20, 47, 111, 124, 130, 134.)

Torksey (183), 7 miles south of Gainsborough, was once an important port at the junction of the Fossdyke and the Trent, and the probable scene of the baptism by Paulinus. The so-called "castle," built in the late sixteenth century, was the Hall of the Jermyn family, and being occupied by the Parliamentarians was captured and burnt by the Royalists from Newark. Torksey has given its name to china and pottery made here. (pp. 82, 97, 99, 100, 116, 143, 153.)

Wainfleet-All-Saints (1258), a market town and port 15 miles north-east of Boston, on the Steeping river and Wainfleet Haven, which reaches the sea at Gibraltar Point, 5 miles away. It is chiefly notable as the birthplace of William Patten (or Barbour), best known as William of Waynflete, who was Provost of Eton, Bishop of Winchester, and Founder of Magdalen College, Oxford, in whose chapel is the tomb of his father, which used to be here. The Bishop also built the existing school at Wainfleet of red brick, much resembling Tattershall Castle in appearance, but about forty years later in date. (pp. 47, 96, 106, 140.)

Whaplode (2270) is a large village on the Holbeach to Spalding road, on which some of the finest churches of the county are situated, that of Whaplode being among the first rank. There are seven bays in the nave, the four eastern ones and the chancel arch well-developed Norman, the three western Transitional, the tower (on the site of the south transept), lowest stage Transitional, second and third Early English, fourth Early Decorated. (p. 120.)

Winteringham (606), once a market and corporate town, seven and a half miles west of Barton-on-Humber, is close to the north ending of the Ermine Street, at Flashmire (Ad Abum). Kirke White, the poet, was for a time a pupil here at the old Rectory. Here St Etheldreda is supposed to have landed, when she was fleeing from her second husband, en route for Ely, and at West Halton close by, where she stopped some time, she is said to have founded a church which is dedicated to her. The church has Transitional Norman piers in the nave, with semi-circular arches on the south and pointed ones on the north, an Early English chancel, and a Perpendicular tower. (pp. 10, 17, 42.)

Winterton (1426), a small market town on the Ermine Street two and a half miles south of Winteringham. Both towns derived their names from the tribe of Winterings. An eagle of a Roman Legion was found here, and close by several Roman pavements, which were figured by William Fowler, who lived here. There is a fine cruciform church, with a pre-Norman tower, its upper portion Early English, an Early English nave, and a Decorated chancel. The altar-piece is by Raphael Mengs. (p. 115.)

Woodhall Spa (1484) is an inland watering-place 3 miles south-west of Horncastle, of considerable and fast rising importance in the treatment of gout, rheumatism, and scrofula, due to the exceeding richness of the natural mineral water in salts of Iodine and Bromine. The spring was accidentally discovered about 100 years ago in the course of an unsuccessful boring 1000 feet deep for coal. The water flows through a soft spongy rock at a depth of 540 feet, and yields from 16,000 to 20,000 gallons a day. There is a spacious pump room and bath establishment of the most modern character. Woodhall is situated on a sandy

The Church Walk: Woodhall Spa

and gravelly soil, and is in the midst of extensive woods. It is sheltered on the north and east by the Wolds, and within easy reach are Tattershall Castle, Scrivelsby, Somersby, Revesby Abbey, and Kirkstead. (pp. 28, 40.)

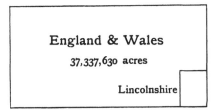

Fig. 1. Area of Lincolnshire (1,705,293 acres) compared
with that of England and Wales

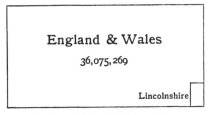

Fig. 2. Population of Lincolnshire (563,960) compared
with that of England and Wales in 1911

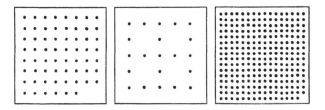

England and Wales 618   Lincolnshire 212   Lancashire 2550

Fig. 3. Comparative Density of Population to the
square mile in 1911

(Each dot represents 10 persons)

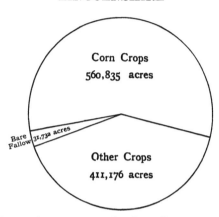

Fig. 4. Proportionate area under Corn Crops compared with
that of other Cultivated Land in Lincolnshire in 1911

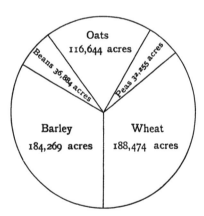

Fig. 5   Proportionate area of chief Corn Crops in
Lincolnshire in 1911

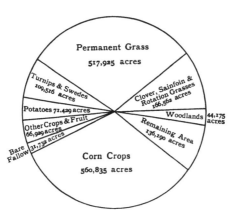

Fig. 6. Proportionate area of land in Lincolnshire
in 1911

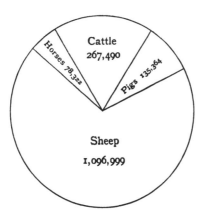

Fig. 7. Proportionate numbers of Horses, Cattle, Sheep,
and Pigs in Lincolnshire in 1911